A WORLD TO LIVE IN

A WORLD TO LIVE IN

AN ECOLOGIST'S VISION FOR A PLUNDERED PLANET

GEORGE M. WOODWELL

THE MIT PRESS   CAMBRIDGE, MASSACHUSETTS   LONDON, ENGLAND

This book was set in Stone Sans and Copperplate by the MIT Press. Printed and bound in the United States of America.

ISBN: 978-0-262-03407-4
Library of Congress Cataloging-in-Publication Data
Names: Woodwell, G. M.
Title: A world to live in : an ecologist's vision for a plundered planet / George M. Woodwell.
Description: Cambridge, MA : The MIT Press, [2015] | Includes bibliographical references and index.
Identifiers: LCCN 2015038373 | ISBN 9780262034074 (hardcover : alk. paper)
Subjects: LCSH: Restoration ecology. | Global environmental change. | Global warming. | Environmental degradation. | Pollution. | Air--Pollution. | Radioactive pollution of the atmosphere.
Classification: LCC QH541.15.R45 W65 2015 | DDC 577.27--dc23 LC record available at http://lccn.loc.gov/2015038373

10 9 8 7 6 5 4 3 2 1

CONTENTS

PREFACE: THE POSTINDUSTRIAL EARTH   VII

ACKNOWLEDGMENTS   XV

PART I: LIFE ON THE SKIN OF THE EARTH

1   IN THE BEGINNING   3

2   NUCLEAR ENERGY: THE COMMONS REDEFINED   11

3   DDT DRIVES A GEOCHEMICAL TEMPEST   29

4   CARBON AND THE CLIMATIC DISRUPTION   49

PART II: ENVIRONMENT IS POLITICAL: CLIMATE HEATS UP

5   THE GLOBAL COMMONS: A CORPORATE FEEDLOT   67

6   CLIMATE IN THE TIDES OF POLITICS   89

7   THE ADAPTATION MYTH: AN ATTRACTIVE CONCEIT
       THREATENS ALL   103

8   THE LIMITS OF BIODIVERSITY   119

PART III: WHICH WORLD?

9   GOVERNMENTS: WHITHER IN THE STORM?   137

10   A NEW DEPARTURE   155

11   *SIC UTERE:* A WORLD TO LIVE IN   173

NOTES   193

INDEX   219

The seventy or more years that have passed since the dawn of the nuclear age have brought greater changes to the earth than any similar period in all of human history. The most fundamental involve intrusions into the ecology of the globe, the biosphere, whose life-giving envelope of light and air and climate has been changed irreparably in a century-long storm of industrial development—a wink in the half-billion years of biotic evolution. The changes have been legion: many threatening, some promising. They have touched the lives of all, crushed some and empowered others. They have entrained changes in government that undermine democracy and civil rights, produced countless innovations in human affairs, and created gross transformations in economic and human prospects.

For an ecologist with an interest in the human condition, ever seeking paths to a stable and equitable future in a world of seven billion humans, and more, the trip has been both fascinating and frightening. I've been fortunate to join in these efforts—a gift of chance, but a rich one that offered unusual experience in science and conservation and government over those years. I have attempted to capture some of that experience in this book, not to write a memoir, but to offer a scientist's perspective on how we have come to this pass, the shoals we must now navigate, and how we might find a far brighter future than now appears on the horizon. As we approach the third decade of the new millennium, the shadow of a poisoned earth and infinitely destructive climatic disruption threaten every element of what we might hope for as a human future.

I have not been alone in this mission. I have had the benefit all along the way of rich associations with scholarly colleagues of great talent, many of whom have written at length on visions of the future, and some of whom, such as Lester Brown, Amory Lovins, James Gustave Speth, John Adams,

Richard Ayres, John Holdren, and Paul Ehrlich, have built institutions to explore the issues with ever-greater cogency.[1] Not surprisingly, although each follows a different path, their perspectives fuse around the universal need to preserve the globe's biotic systems as the sine qua non of success in passing a civilization and a habitable earth on to successive generations. And that, of course, is my topic, formulated by colleagues in various contexts but by none better than Speth in *The Bridge at the Edge of the World*:

> The escalating processes of climate disruption, biotic impoverishment, and toxification that continue despite decades of warnings and earnest effort constitute a severe indictment, but an indictment of what, exactly? If we want to reverse today's destructive trends, forestall further and greater losses, and leave a bountiful world for our children and grandchildren, we must return to fundamentals and seek to understand both the underlying forces driving such destructive trends and the economic and political system that gives these forces free rein. Then we can ask what can be done to change the system.[2]

A youthful attachment to an heirloom farm in a far corner of the town of York in southern Maine led me early in life to question the ways in which human welfare and wealth were tied to land and soil, and the environment more generally. That part of Maine is hilly drumlin country, with diversified land, fruitful enough for its farmers to have made a living on in the first centuries of European settlement. Successive families on this old farm over three hundred years and more had found they could do just a bit better than mere subsistence. While the more recent ancestors among them forged their own tools and grew celery in the swamp, their potential was limited by the agricultural potential of land, locale, and later, economic shifts, technological change, and other factors quite beyond their control. Subsistence ultimately required more than the land, at least more than that land, could offer. The economic crash of 1929 stole family savings through the failure of the local bank. The house burned in 1933 and was not then rebuilt. Full-time farming ceased—a common occurrence in those years and later as the locus of commerce in farming moved westward to some of the richest soils in the world, and farming itself was transformed into an increasingly mechanized industry on more easily managed landscapes.

That was the story as commonly written and understood. But it was not quite that simple. Economic and political changes over time brought a consolidation of power in corporations that could take advantage of the farmer's ties to land and crop and the need for an immediate market for milk and other perishable farm products. Corporate monopolies could easily manipulate prices, and did. Ties to life-giving land passed on for generations became bondage, with commitments to markets developing into a new form of tyranny. Partial remedies lay in laws protecting farmers and their markets. "Milk-control boards" begun in some states in the 1940s, for example, helped assure a market and a fair price to the farmer, although industrial interests entered into all such adjustments as well. As corporate forces grew more powerful and spread across the nation, it became increasingly clear that preservation of human welfare was not a simple product of wise management of intrinsic resources of land and water in a free market. The market was not free. A fair and vigorous economic system required regulation by a government capable of, and interested in, protecting the welfare of individuals—all of them.

The intensification inherent in both population growth and industrial development worldwide has put soaring new demands on all resources. Intensification has always accompanied civilization's growth, but it has been especially conspicuous in the seven decades since the close of World War II, first in the United States and other Western nations, and more recently in China and other emerging economies. The global population has expanded over those decades from about 2.3 billion people in 1940 to more than 7 billion today, and continues to rise. In that period, world economic activity as measured by the gross domestic product (GDP) soared more than twenty-five-fold from about $3 trillion to about $85 trillion in 2012.[3]

That growth has involved the development of powerful economic institutions—in some nations, state organized, and in others, such as the United States, mainly in the form of private sector banks and corporations—whose financial influence entrains political power. Government is easily bent to favor corporate interests; human rights and public welfare usually have less priority. Early in that evolution the competition was mainly for control of agricultural markets, but it quickly became a competition for land, water, energy, locale, and people as opportunities for further growth diminished.[4]

Weak or nonexistent governmental controls combined with powerful corporate competition for greater profits and continued economic growth have led to the acceptance of corporate industrial wastes as common property. Wastes were dumped without thought or serious objection into the public realm at no cost. It mattered not at all whether the wastes were solid materials, industrial chemicals, noxious gases, or relatively benign carbon dioxide. The practices started early in the competitive game of economic development and were presumed by many to be a corporate right—the basis of profits and jobs. Early protests against these invasions of the public sphere devouring air, water, and even life itself were deflected with claims that such practices were necessary for the financial profits that would bring economic advantages to all. The worldwide economic depression of the 1930s had hammered that concept into the public mind: businesses were to be protected at all costs.

To be sure, there were those who warned of the consequences of rapid population growth and the kind of industrialization taking place, and asserted that the earth had finite limits to what life it could sustain in the long run. A flood of books had emerged by midcentury beginning with John Steinbeck's 1939 classic, *Grapes of Wrath*, Paul Sears's 1935 *Deserts on the March*, William Vogt's *Road to Survival* (1948), Fairfield Osborn's *Our Plundered Planet* (1948), and Aldo Leopold's *A Sand County Almanac* (1949). Later, in the 1960s and 1970s, powerful additional statements emerged from Rachel Carson (*Silent Spring*, 1962) and Paul Ehrlich (*The Population Bomb*, 1968), among others. James Lovelock in 1974 introduced the idea that the living earth, which he called "Gaia," has made itself, and in the long run of thousands to hundreds of thousands or more years, well beyond most human horizons, has the potential for restoring itself.

One of the most telling of these analyses—often cited but whose full import is seldom appreciated—was that of an admirable friend and philosopher, Garrett Hardin, in his 1968 article titled "The Tragedy of the Commons."[5]

He pointed to the advantage claimed by a farmer's pasturing an additional animal in a common lot. The owner of the stock claims the full profit while the damage from the grazing is shared among all those using the pasture. There is a clear economic advantage to the individual in exploiting the commons in the short term despite its progressive decline. Protection of the commons demands mutual efforts mutually agreed to. The lesson is powerful, although it is occasionally inverted to argue that the market is enough to set prices that will protect the commons. Hardin in fact defined the core purpose of government as protecting the public welfare, the human birthright to clean air, water, and a place to live, among other rights. The welfare of each is to be protected from the depredations of all, and the welfare of all is to be protected from the depredations of each. Governments are established with the express purpose of providing such rules and enforcing them, corporate interests and profits notwithstanding. Any defense of the environment depends on that principle applied at all levels, local, national, and international.

As the human enterprise has grown to push against or exceed global bio-physical limits, the rules—and even more, the practices—of the nations in preserving human rights and dignity have not kept pace. Corporate inter-ests have been able to control or enthusiastically assume governmental roles in "public welfare," even to the point of redefining it to favor their own interests and line their own pockets. There are many consequences of these developments, but the two that now count most heavily for our future are climatic disruption and the corruption of the chemistry of the global environment. Both undermine all life on earth and subvert human welfare in the most fundamental of ways.

Conventional economics and governmental practices are challenged now in ways never before contemplated. At issue is the intent and capacity of governments—and all of us as citizens—to protect both the public sphere and the integrity of the biosphere, the basic conditions that support all life, the sine qua non of the continued viability and progress of civilization.

For the ecologist minted in the late 1950s, the agenda outlined above wrote itself. Suddenly, environmental issues had a global side. The Bomb did that, and the change in perspective was indelible. Now more than half a century later, we have learned to think of globalization as the process by which trade and commerce have increased the world over, along with the internationalization of finance, communication, and rapid travel. While all these facets of modern life have increased enormously in recent decades, so has the earlier dimension of globalization, the chemical penetration and pollution—the poisoning of the planet, from nuclear radiation to pesticides, to carbon dioxide and methane, to a thousand other industrial wastes. These have come with a growing appreciation by scientists and lay-persons alike that we all live on one globe, the integrity of the biosphere as a whole is critical, there's fragility to the conditions that make life possible, what we do as a society affects people everywhere in profound ways, and what they do elsewhere affects us.

I came to ecology at the turning point when the struggles to recognize the integrity and continuity of the biosphere as essential to the human future became critical. I was fortunate in being able to join in the research and to find my way into national and international efforts, both governmental and nongovernmental, to see those new insights put into public policies. The transition from insight to policy lags, even at this late date approaching the third decade of the new millennium. But the reality of the needs for insights and bold action only become clearer minute by minute, as outlined in this brief book.

* * *

In part I, I draw on the perspectives developed by Charles Darwin and George Perkins Marsh of the earth as a living system, an evolutionary product vulnerable to disruption in human hands. The experience with nuclear weapons offered strikingly new insights globally into what was likely to follow as the economic system expanded rapidly under lax regulation. Early research into nuclear radiation, in which I was involved, showed some of the potential dangers and the need for new models of closed industrial systems. Experience with DDT revealed the downsides of industrial intrusions, and the fierce resistance of corporate agricultural and commercial interests when their license to profit from pollution was threatened. So, too, with the wastes of the fossil-fuel industries that poison the global atmosphere and can change, or may have already changed, global climates sufficiently to end this civilization.

In part II, I discuss major barriers to corrective action in building a world safe for all life, including people. The largest barriers are the free market ideology and the license it gives to commerce to exploit the commons as a "corporate feedlot." Polarized politics driven largely by the profit motive has nullified international agreements, and frustrated national needs and constructive moves toward a poison-free world. Then I explore conservation efforts as well as the need to look beyond the preservation of species to the protection of the natural communities that are their habitat. I explain some of the controversies encountered along with the shameful resistance demonstrated when the US Congress rejected the "US Biological Survey." Willful neglect of the accumulated wastes of the fossil-fueled industry is now the most dramatic illustration of environmental disruption. I bring science to an understanding of this crisis, and the attractive but dangerously false assumption that it is possible for industrial civilization to continue on substantially the same path with fossil fuels and simply adapt to the climatic disruption. The assumption that we can "muddle through" is a ticket to catastrophe.

Part III offers paths to a solution. All paths require a restoration of respect for the full range of life on earth and its preservation. Biotic processes have a greater degree of control over the composition of the atmosphere in days to months than any other factor and will be key in any solution. In addition to abandoning fossil fuels, we must build a new world of essentially "closed" industrial systems that operate within the limits set by the local and regional natural communities. Climates can be stabilized now, and the world set on a course to return the atmosphere over a century to approximately the heat-trapping gas content of 1900, but only through a concerted and worldwide effort. It will require not a slavish reliance on the free market system but instead a recognition of the role of government in preserving, as an inviolable human right, the physical, chemical, and biotic integrity of a finite biosphere. We have the scientific insights, the experience required, and the political and economic systems to build an exciting, renewable human environment. Required is the political dream based on the recognition that the global commons, the earth's living systems, are the heritage of all humankind. Their restoration and protection reverse current trends and open an inspiring series of new paths to richly rewarding lives in an infinitely renewable civilization.

ACKNOWLEDGMENTS

In writing this book, I have enjoyed the interest and generous support of the Woods Hole Research Center, and subsequently, after retirement from the Center, the Natural Resources Defense Council, where I have been a board member for more than four decades and, most recently, affiliated as a Distinguished Scientist. I have been encouraged in particular by longtime friend and associate Merloyd Lawrence, whose patience, confidence, and advice have been continuous, ever penetrating, and delightful. She led me to editor Jonathan Cobb, who leaped to the challenge and provided the well-informed skeptical reader any author must please. He worked hard to dilute hyperbole, soften inherent dogmatism, and eliminate repetition, all of which had seemed necessary to hold the interest of writer and reader, and which despite his resilience, effervescent wisdom, and restraint, oozed out anyway.

I have been extraordinarily fortunate in having had rich experience with distinguished colleagues and teachers in a diversity of academic, governmental and nongovernmental institutions over decades. These individuals reach from Dartmouth and Duke, to Yale and the University of Maine, to the US Navy, Brookhaven National Laboratory, Marine Biological Laboratory, Conservation Foundation, World Wildlife Fund, World Resources Institute, Ocean Conservancy, National Academy of Sciences, and National Research Council as well as scientific organizations such as the Ecological Society of America and scores of institutions I have visited around the world as a guest over many years.

I owe particular personal debts in my training to John Adams, Gus Speth, Dean Abrahamson, Dick Ayres, David Hawkins, Larry Rockefeller, Linda Greer, Ashok Gupta, and many other staff members and trustees of the National Resources Defense Council; Kilaparti Ramakrishna, Richard Houghton, Foster Brown, Tom Stone, Greg Fiske, Toby McGrath, Dan Nepstad, Michael Ernst, Allison White, John Holdren, Greg Fiske, Dennis Dinan, Mary Lou Montgomery, Eric Davidson, Tom Lovejoy, Steve Curwood, Iris Fanger, John Cantlon, M. S. Swaminathan, and Larry Huntington, all longtime staff members or trustees of the Woods Hole Research Center; Jerry Melillo, John Hobbie, and Bruce Peterson of the Marine Biological Laboratory's Ecosystems Center; and decades of friendship and discussions with Herb Bormann, Paul Ehrlich, Al Gustafson, Bill Schlesinger, Robert Howarth, Sandra Steingraber, David Orr, Bill McDonough, Jim MacNeill, Ola Ullsten, and scores of other friends and colleagues around the world.

But the most powerful teachers of all have been the in-house crew of critics and loyal agents of humor and help: Caroline, Marjorie, Jane, and John Woodwell, academics and scholars all, who dwell in far-flung places but return to lend interest, criticism, and support to such matters as the great issues of life and times, while Katharine, with smiles and few words, towering grace and talent, keeps a sharp eye and loving hand on the helm.

I   LIFE ON THE SKIN OF THE EARTH

There is grandeur in this view of life ... whilst this planet has gone cycling on....
[F]rom so simple a beginning endless forms so beautiful and so wonderful have
been, and are being evolved.

—Charles Darwin

He was born in 1809, the year of Abraham Lincoln's birth, into a family of
scholarly tradition in England. He was blessed with an insatiable curiosity
that fed a retentive mind. After college, he escaped a career in medicine
and later the clergy by becoming a naturalist. In 1831, at age twenty-two,
Charles Darwin joined the HMS *Beagle* officially as a naturalist on its five-
year exploratory cruise. A whole career later in 1859, the threshold of the
US Civil War, Darwin published *On the Origin of Species*.

As the Civil War brought an end to slavery in the United States and fueled
a surging evolution of civil rights that pulsed through a century and a half of
development of the modern industrial state, so the publication of Darwin's
insights into the origins of life shook the world and ultimately became the
foundations of modern biology, including details of the evolution, structure,
and functional integrity of what we have come to call the "biosphere."[1]

The roots of ecology—insights into the structure of the living world—
extend, of course, back well before Darwin described the dimly envisioned
groping of our ancestors over ten thousand years on four continents as they
found their way into agriculture and the domestication of both plants and
animals. As what became the great grain crops were selected and cultivated
in the Middle East, so were the potatoes of Peru and corn (Zea) and cucur-
bits of the Americas turned into cultigens by the persistent, wise farmers
of what became known as the New World.[2] It took not only innumerable
observations that like begets like and that the selective breeding of plants

and animals can favor human interests but also the global experience and astute observations of Darwin and Alfred Russel Wallace in the nineteenth century to envision competitive selection as a natural process capable of developing an enduring, living world. And even then the vision was incomplete and crude as Darwin advanced it, almost as a mere suggestion, in 1859 in the final paragraphs of *On the Origin of Species*:

> It is interesting to contemplate an entangled bank, clothed with many plants of many kinds, with birds singing on the bushes, with various insects flitting about, and with worms crawling through the damp earth, and to reflect that these elaborately constructed forms, so different from each other, and dependent on each other in so complex a manner, have all been produced by laws acting around us.... There is grandeur in this view of life, with its several powers.[3]

That thought served as the foreword to a torrent of studies and treatises by other gifted observers and analysts on evolution and ecology over the next century and a half, and continuing today. These writings underlie contemporary ecology, conservation, and governmental laws designed to protect air, water, and land in the interest of all.

Even as Darwin was defending his insights into how the living world was built, the American George Perkins Marsh in 1864 was observing in his famous book *Man and Nature* that the structure of that world was already threatened by the magnitude of human activity. He wrote

> [about] the changes produced by human action in ... the globe we inhabit; to point out the dangers of imprudence and the necessity of caution in all operations which, on a large scale, interfere with the spontaneous arrangements of the organic and the inorganic world; to suggest the possibility and the importance of the restoration of disturbed harmonies and the material improvement of waste and exhausted regions; and, incidentally, to illustrate the doctrine that man is, in both kind and degree, a power of a higher order than any of the other forms of ... life, which, like him, are nourished at the table of bounteous nature....
>
> The action of man upon the organic world tends to derange its original balances, and while it reduces the number of some species, or even extirpates them altogether, it multiplies other forms of animal and vegetable life.[4]

Now, despite a century and a half of insights into global ecology, and abundant warnings about the enormous expansion of human numbers and intensification of industrial development, we fool ourselves into managing our affairs as though the world were infinitely resilient. I was no different from my peers, waltzing into a world already impoverished, accepting both the status quo and the ways of the world at that moment, and at least initially, adopting without question the standards and methods of the time and culture. Had I been led or become personally inclined to enter business or finance, an early interest in the basis of wealth and human welfare on an agrarian landscape might have followed a more conventional track. But I discovered at Dartmouth in the immediate postwar years a subculture of science that inspired me and seemed to reach to fundamentals as to how the world works. Over the intervening years that subculture has exploded to challenge the prevailing worldviews with history and insight that reveal the frightening assaults we as a species have engendered, still under the assumption that the world is large and resilient. These assaults are marked now in scores of ways—among them in my own adult life, the poisoning of the Gulf of Mexico with agricultural wastes and oil, the contamination of the global atmosphere with radioactivity from nuclear weapons and from accidents, and the atmosphere's poisoning with mercury, oxides of sulfur, other industrial wastes, and heat-trapping gases, each a massive intrusion into the biosphere, each as though it is an innocent, independent anomaly, an error that is easily corrected by natural processes and soon to be forgotten.

They are not anomalies, and they are forgotten at great hazard. They are callous, repeated blows against a community of survivors of the same selective forces that Darwin envisioned as building species. And they leave scars in the matrix of life that reduce its capacity to support all life. They are the routine but erratic sequelae of the human scramble for natural resources, especially the scramble for energy in support of currently popular, if misleading, economic and political ambitions. And they are not innocent. They are intrinsic, inevitable, and predictable failures of a corporate culture based on competitive greed supported by governments. Nor are they corrected quickly, or in some cases at all. They simply pass into history, and we live with the result: an earth incrementally and progressively impoverished as it erodes under the burden of billions of humans struggling to thrive, and an industrial worldview that recognizes no limits on its expansion and intrusions. This is the worldview at the moment presented to each new citizen at birth.

* * *

Civilization requires for its survival a triad of successfully functional systems: political, economic, and environmental. At its best, the *political system* develops and defends the rules protecting human rights and welfare;

the *economic system* enables the reliable production of vital goods and services along with their equitable exchange; and the *environmental system* gives us a biosphere that predictably assures the resources essential to life. As we see over and over again in the news of the day, immediate issues of politics and economics draw continuous attention, often to the neglect and even detriment of the environment. Yet failure in any one segment of this triumvirate can doom the whole.

Contemporary failures in economics and politics are many, even within the US system, which had been until recently one of the most sensitive and successful in the world. That commanding position has been lost. Chris Hedges and Joe Sacco in *Days of Destruction, Days of Revolt* present a devastating analysis of current socioeconomic conditions in the United States, showing frightening trends toward increased poverty and economic disarray. The authors focus on the people affected in geographic regions that have been abused and then abandoned by corporate activities, such as the mined-out mountains and debris-filled valleys of the southern Appalachians as well as areas of southern Florida where the industrial production of large-market cash crops has been allowed to destroy formerly stable human communities. The environment is almost an incidental victim of these depredations, but its impoverishment becomes a further cause of pathos and misery.[5] Corporate interests in these and other regions have displaced responsible government, civil rights, the public welfare, and the prospects for an attractive and wholesome human future.

We live now in a world in which the global environment is not simply at risk; it is eroding with alarming rapidity. That erosion encompasses not only the accumulation of the sacrificial zones and communities Hedges and Sacco identify and deplore but also the far more comprehensive global changes in climate and environmental chemistry now sweeping the earth. Corporate greed is one, but not the only, cause, and corporate political power has been exercised regularly in the United States and abroad to frustrate efforts to reverse the trends. A second major force is the push for growth, the expansion of human numbers, and understandable aspirations of all for richer lives as industrial development and commerce spread globally.

The intensifying dilemma of increasing demands on finite resources is the "tragedy of the commons" so brilliantly articulated in the famous 1968 paper by Garrett Hardin mentioned in the preface.[6] Hardin, who wrote during the mid-twentieth-century surge in environmental concerns, showed that individuals find a personal, immediate advantage in adding ever one more head to graze a pasture shared by all. So it is among corporate and even governmental interests exploiting other common property, such as air, water, and land. Each claims a private profit while the costs are shared among the public at large. Unless rules of use can be established and mutually respected, the resource is degraded, benefits are inequitably distributed, and ultimately all suffer. We establish governments to define and enforce such rules for the protection of all.

The issue has become acute globally as the frequency and severity of disasters linked to fossil fuels, energy, and the climatic disruption have increased. Consider, for example, the northern summer 2010 through spring 2011—the period in which a huge BP oil spill occurred in the Gulf of Mexico, and nuclear reactors melted down at Fukushima, Japan, in the wake of an earthquake and a resulting tsunami. It was also the latest of successive years of unprecedented summer heat in the Northern Hemisphere, offering a glimpse of summers to come as the earth continues to warm under an increasing envelope of heat-trapping gases. It was a year of giant storms in the United States, with tornadoes ripping towns apart as far north as Massachusetts and Hurricane Irene's winds tormenting the East Coast from Florida to Montreal. And while extraordinary rains from Irene and its immediate sequel, Lee, flooded regions from Pennsylvania to Vermont, drought intensified in the Southwest. Texas burned first, then months later, New Mexico and Arizona. Later, in spring 2012, sections of Colorado and Utah were consumed in unprecedented fires fed by drought and hot, dry winds. Thousands of suburban or rural forest dwellers were displaced when many houses were lost.

Neglected in the mainstream news reports at the time was any suggestion that the severity of the storms, rains, floods, intensifying drought, and fires were indicative of a fossil-fuel-driven climatic disruption. Only when a year later, in October 2012, Hurricane Sandy brought disaster to the New Jersey and New York shores did some of the mainstream media presentations acknowledge the role that climatic disruption could be playing. The storm surge flooded a segment of the New York City subway system, and brought home to thousands the reality of sea level rise along with its potential for introducing long-term inconvenience and great additional expense into life in a large city. These disruptions are but the beginning, the front edge, of the global changes in climate now gripping the earth as a whole even if any particular event cannot be tied with certainty to climatic change alone.

Similar disasters are now becoming part of life on every continent. The human costs accumulate quite outside the realm of the extraordinary profits reaped by corporations engaged in mining and selling the oil, coal, and gas resources that are part of our collective heritage. They reveal the stark ignorance—willful or otherwise—of, for example, ExxonMobil Corporation, whose ultimate products, the wastes of the industry, accumulate in the atmosphere as heat-trapping gases, dumped into the public realm without immediate cost or apparent serious concern that they poison all the earth. So powerful are the financial interests thus engaged that governments, even successive administrations over more than twenty years in the United States, have been unable or unwilling to bring this colossal corrupting system under control. A citizen must wonder, just what has happened to the overriding purpose of government to protect and secure a future for all of us?

The US Congress, ever pushed by conservative financial interests, further slipped during these years into a state of ideological denial as one political party succumbed to the myths and wealth of the corporations engaged in highly profitable fossil-fuel and related businesses. A strangely complacent and compliant Congress, not immune to the flow of campaign money, and in denial of its responsibility to protect human welfare, has in effect chosen to deny fundamental facts of nature. James Inhofe, a Republican senator from Oklahoma, solidified his notorious intransigence in a book published in 2012, *The Greatest Hoax: How the Global Warming Conspiracy Threatens Your Future*.[7] Joe Romm, who writes regularly on climatic disruption, pointed out that as Inhofe announced his book's publication, his home state was experiencing, according to the Oklahoma Climatological Survey, "the hottest summer of any state since records began in 1895 with a statewide average of 86.9 degrees F."[8] Meanwhile, vociferous Republican majorities in both the US House of Representatives and Senate in 2015 also repeatedly denied the disruption of climate globally by the human-caused warming of the earth. This majority appeared to consider such environmental disruptions, especially including the climatic one, as a simple matter of opinion, easily set aside as a political inconvenience, a bothersome whim of impractical dreamers. Worse, both houses of Congress appeared to wish to rescind previously developed and celebrated environmental regulations, the rules that protect essential resources such as air and water from despoliation by divers commercialism, and that if more carefully written and even more closely observed, would have protected us from any such intrusion as those mentioned.

Looking back over the last several thousand years of human history, the Australian scientist Anthony J. McMichael, in a remarkable review of the role of climate in the rise and fall of civilizations, offered a disturbing appraisal: "The fact that drought has been the dominant historical cause of hunger, starvation, and consequent death casts an ominous shadow over this coming century, for which climate modeling consistently projects an increase in the range, frequency, and intensity of droughts." In measured terms he went on: "Compared with the historical record [the projections are for] ... an extreme and rapidly evolving long-term change ... without precedent during the Holocene [approximately the last ten thousand years since the last glacial retreat].... Such a change will surely pose serious risks to human health and survival, impinging unevenly, but sparing no population."[9]

Lester Brown, an agricultural economist, shares the alarm, but emphasizes the threats posed by growth in human numbers on the demands for food under such threatening climatic circumstances.[10] His perspective builds on the doubts that the British cleric Thomas Robert Malthus expressed in the late eighteenth century in his now-famous essay on population.[11] Malthus observed that the potential for growth in the number of people is exponential, and unchecked, eventually exceeds the potential for growth in resources, particularly in the area available for agricultural land and food production. It was not until the latter decades of the twentieth century and the first decade of the twenty-first, however, that a consensus began to form among environmental scientists that the human expansion, which passed six billion in October 1999 according to a UN study, was increasingly challenging the biophysical limits of the earth.[12] The most compelling observation, however, was that of the then-conspicuous climatic disruption.

In centuries past, it seemed reasonable to envision political and economic systems without defining details of the environmental systems supporting them. Now, as the intensity of requirements for resources of air, water, and land increase, the three realms, government, economics and environment, have become more closely knit and interdependent. The interests of industrial leaders to amass profits by spreading the costs, environmental and otherwise, as widely as possible among the public has become an increasingly unacceptable burden, even as it is vigorously defended by its beneficiaries, who are often able and willing to buy political support through paid lobbyists and political campaign contributions. The corruption of the political as well as biophysical commons is a further amplification of the "tragedy" that Hardin so clearly set forth; it has now become a seriously destructive global issue—possibly the great stumble, the ultimate disaster, of this generation.[13]

* * *

Darwin, Marsh, and Lincoln were nineteenth-century contemporaries, and each dealt with a great issue of his time that emerged as a great issue of our time. From their lives, fully as transformative in public affairs as the cotton gin and the steam engine were for industry in those years, have emerged extraordinary advances in biology, medicine, agriculture, land use, and ecology as well as perspectives on civil rights, government, and on equity and justice in human affairs. The transitions have come slowly, sometimes from strange directions. The twentieth century saw the rise of numerous voices, Hardin's among them, calling attention to the ill logic of continued population growth and unregulated but expanding demands on common property resources. The demands have only intensified in our century and now require urgent action. Can we use, as I believe we must, the chain of critical environmental experiences of recent decades to define a course that will restore environmental security and preserve it over the next decades? Or must we suffer the multiple catastrophes entrained in the current trends?

The most alarming of all man's assaults upon the environment is the contamination of air, earth, rivers and sea with dangerous and even lethal materials. This pollution is for the most part irrecoverable, the chain of evil it initiates not only in the world that must support life but in living tissues is for the most part irreversible.

—Rachel Carson, *Silent Spring*

It was July 1, 1946, less than a year after the final stroke of World War II had destroyed two Japanese cities. The bombs had substantially sterilized two sizable segments of the earth, each within minutes, each deliberately, each with one atomic bomb that left an indelible mark, a scar on the biosphere and in the human consciousness beyond repair. For these two cities, a half-billion years of biotic evolution had been all but extinguished.

A large new military venture was under way in the South Pacific to further "test" the effects of these new weapons. The target was Bikini Atoll, selected because it was remote from large population centers and the inhabitants, despite understandable reluctance, could be forced to move without serious international objections. It was only the beginning of a series of nuclear blasts that continued in the Marshall Islands for more than a decade, and targeted not only Bikini but also Kwajalein, Eniwetok, and in 1954 by mistake, Rongelap, when the much larger H-bomb was used. Meanwhile, tests were in progress at the Nevada Test Site east of Los Angeles, and proposals were advanced somewhat later to use nuclear blasts to excavate a new sea level canal across Colombia and a harbor at Cape Thompson on Alaska's northwest coast—part of the wishful thinking embraced in a program called, with a biblical allusion, Plowshare.[1]

And the United States was not alone. The USSR, France, and later China joined in the competition to advertise their triumphs with tests in the atmosphere. Most of the atmospheric detonations ceased in 1963 with the signing of the Limited Test Ban Treaty in Moscow. Then the game went underground. By then the earth as a whole, both hemispheres, had been contaminated with radioactivity that was easily traced, and in its tracing, revealed details of food webs and chemical trails never seen previously. While the military insights gleaned from the scores of tests may have been important, far more compelling was the evidence of the potential for subtly destructive human intrusions into the biosphere.

Foremost, of course, were the brutal effects of the blasts.

David Bradley, a physician, recounted the report of an observer sent to aid the Japanese in Nagasaki, the second of the two cities destroyed. Bradley wrote:

> A friend of mine, a navy surgeon, ... was ordered to see what remained of housing, water, electricity, transportation, communications ... for the occupying troops ... six weeks after the bombing.... The captain's gig landed them on a float at the bottom of a tall ship's wharf. A staircase of cement steps led to the top ... [where] the doctor found the whole city opening out. Only it wasn't a city; it was nothing. For nearly two miles a plain of ash and rubble stretched away before him to some sharp green foothills. In the distance one sturdy lump of a building still stood, on a knoll beside the squashed steel webwork of the Misubishi torpedo plant.
>
> Nothing moved. Not a person. Not a child running or a dog scrounging. No sounds. Only the lapping of the morning waves against the boat landing below.[2]

Six weeks after the bombing.

A world devoid of life.

Few have seen, or wish to contemplate, such a world. At the time we saw the devastation as the special new gift of war, not realizing the potential of the explosive growth-to-come of the corporate industrial world for equivalent destruction. And then there was the global contamination of the biosphere, especially heavy in the Northern Hemisphere, caused by the debris from nuclear detonations. The contamination gave surprising and rather frightening insights into the spread and effects of radioactivity from the weapons. The debris circulated, first in the high atmosphere, the stratosphere, thrust to those great heights by the blast itself. Later it was mixed into the lower atmosphere and rained out to enter the earth's intricate food webs.[3] The insights were new, astonishing glimpses into aspects of nature that we had only guessed at previously. I realized but slowly that these insights were the bellwether of a flood of industrial wastes and problems on a global scale already festering and yet to emerge.

For two decades following the Second World War it mattered not whether it was a Polynesian breakfast, a Siberian prisoner's meager dinner, or *haut couture de Paris*—all carried poisons from the new, esoteric and terrifying technology of nuclear weaponry.

It was, then, a refreshingly optimistic and bold twist after the war to use that new technology hitherto only used in war to boil water. The world needed energy and steam could turn turbines and feed energy into the expanding industrial civilization. Large nuclear reactors designed to generate heat without the danger of nuclear criticality (an uncontrollable fission reaction) seemed to offer almost-unlimited potential in the generation of electricity. A new industry in generating energy gained momentum rapidly in the decades following the war. The first power reactor was commissioned in Obninsk, Russia, in 1954. By 2015, there were 439 nuclear power plants operable in thirty-one countries. Globally 69 more were in some stage of construction.[4]

Safety certainly was of concern from the beginning. Hazards of human exposure to ionizing radiation (energetic radiation capable of making molecules chemically reactive by breaking chemical bonds) were well recognized as cancers, even from exposures low by comparison with the potential releases from bombs or a reactor accident. The new technology carried some absolutes: there could be not only no bombs but also no accidents, no errors in operating reactors, no leaks of radiation from reactor cores, and no leaks from radioactive wastes that had to be isolated essentially forever (two hundred thousand years). Safety demanded that the nuclear industry be a "closed system," complete within itself, with no leaks anywhere in the process of energy production. It required relentless monitoring and perfect maintenance. Yet the potential for contamination with radioactivity once a reactor had been operated for even a short time was huge if the containment were breached in any way. Governmental regulation and inspection were essential.

* * *

The experience with nuclear bombs—whose detonations in the atmosphere were the opposite of closed systems—was still fresh and startling. Everyone realized that the proliferation of weapons of the Cold War and the development of new nuclear power plants opened the possibility of spreading contamination from accidents, no matter how carefully the technology was managed. But the immediate issue in 1960 that triggered interest among scientists and regulators was the surprising discovery by Arnold Sparrow at Brookhaven National Laboratory on Long Island, New York, that plants, at least some of them, were much more sensitive to ionizing radiation than had been previously recognized.[5] The discovery raised new questions about the hazards of nuclear contamination to all life, not to humans alone.

Sparrow was a botanist who had for years been exploring the sensitivity of plants to gamma radiation, which is similar to X-rays, but more powerful. In many of his experiments, he exposed various plants in a garden on the Brookhaven Laboratory site to a large, single cobalt 60 source of ionizing radiation. Surprisingly, pine trees (*Pinus rigida*), nearby but outside the experimental area and only incidentally exposed to low-level radiation, proved far more sensitive than many of the plants in the experiments. In later experiments, under certain circumstances pine seedlings died after approximately the same acute exposures as humans.[6] The surviving mature trees near the experimental field had become malformed and stunted in growth.

Encouraged by the Atomic Energy Commission to study the issue further, Brookhaven Laboratory assembled a small group of scientists to explore how to define the ecological effects of ionizing radiation: what would happen, for instance, if extensive areas of the earth were exposed to levels of ionizing radiation many thousands of times higher than the contemporary background radiation that all life receives?

The issue at that time was a matter of some consequence not only to the Atomic Energy Commission but also to other agencies of government charged with protecting the public health and welfare as well as to the industries busy building reactors. The public, too, would have taken interest if they had been paying close attention to this somewhat esoteric realm of science. The nuclear weapons detonated in 1945 and subsequently had shown that we could contaminate the entire earth and expose virtually every organism on earth to higher than normal risks. In fact we had done so, and governments were heavily involved, working ostensibly in the public interest. At the same time, enthusiasts for nuclear power were spreading effervescent optimism about the "friendly atom," and broadcasting exciting projections about the abundant and inexpensive electric power that the new reactors would bring.

The new nuclear power industry itself, however, was acutely aware of the hazards from a reactor accident. It had sought and received in 1957 a massive subsidy from Congress in the form of the Price-Anderson Act. The act limited the liability of the industry for any accident to $60 million and passed any further liability along to the federal government with an absolute limit of $500 million. (The act was renewed in 1967 and subsequently with higher levels of liability, but the same principle applied.) Commercial insurance was not available at an acceptable cost; only the government's intervention made a nuclear energy industry feasible in the United States. The industry's insistence on the need for the subsidy belied the confidence it showered on the public that the reactors were safe and the industry could operate in a closed-cycle system. Clearly, the public in the vicinity of reactors was vulnerable to a hazard acknowledged by both industry and government.[7] Human health was thought to be the big hazard. Now, with the Sparrow discoveries, it might be more general.

As to ionizing radiation's direct effects on plants and natural communities, there was little experience. The best information came from Sparrow, who had for years been accumulating data on the responses of a wide range of plants to acute and chronic gamma ray exposures. Sparrow's specialty in botany was in cellular structure. He discovered not only that there was an enormous range of sensitivity within the plant kingdom but also, not surprisingly in retrospect, that the most sensitive elements in the cell are the chromosomes. Sensitivity, measured in roentgens of exposure over time (the density and duration of ionizations), was related to the number and size (volume) of chromosomes. Pine trees are vulnerable because they have a small number (twenty-four) of large chromosomes. Organisms such as lichens with a large number of small chromosomes were orders of magnitude more resistant. Some of the breaks in chromosomes caused by exposure to radiation could be seen under the microscope as fragments; others were more subtle, and appeared only as anomalies of growth or form in subsequent generations. All, of course, were mutations (changes in genetic structure). Such damage at the cellular level may kill the cell without killing the organism. If enough cells are damaged, the organism will be affected. Mutations commonly affect reproductive potential, but also tissue and organs if the exposure is high.

Sparrow's work was extraordinary, exciting, and fundamental. He tested ionizing radiation as a mutagen (causing mutations), looking even at the question of whether new mutations might make, for example, crops more productive. One such experiment involved African violets, a particularly good plant for such a study because new plants that are genetically identical can be produced from the leaves of a single plant. He irradiated such a population systematically to produce as many mutations as possible and then allowed the irradiated leaves to develop into plants. Sparrow succeeded in reproducing a range of genetic variability for which the species is well known and admired, but not a single new strain or form that survived.[8] And that has been the history of radiation-produced mutations: they may add to the variability of a species in ways already well established and stabilized, or they are deleterious, cause deformity or premature death, and disappear as the organism dies.

As many others at the time, I was curious about both the potential for power generation and the uses and hazards of nuclear energy. I was then on the staff of the University of Maine in Orono, but I had already decided that I needed much broader experience in ecology than I had yet accumulated. An acquaintance, Eugene P. Odum, was a well-established scientist on the staff of the University of Georgia and a wonderful gentleman. He had published a book on ecology celebrating the "ecosystem"—still an inchoate concept, I thought, even for Odum.[9] But I liked the effort and also his program at the Savannah River Laboratory, an Atomic Energy Commission laboratory with an abundance of radioactive wastes that offered a new realm of potential experience that I considered essential. It was at that moment that F. H. Bormann, a close friend, then at Dartmouth, suggested that I look into a possibility at Brookhaven, also an Atomic Energy Commission laboratory, although focused not on weapons production or the development of reactor power but instead on basic research related to the nuclear age. It was the perfect challenge at the right moment.

The Brookhaven group's initial task was to set up a definitive experiment around the core question of the ecological effects of exposure to high levels of gamma radiation, the principal hazard. I relished the challenge and was pleased not only to join the discussion but also to accept a position on the staff of the biology department to pursue the topic. Suggestions varied widely, even among ecologists. Finally at my insistence we settled on a plan for an experimental exposure of a natural community: a forest on the laboratory site in central Long Island to be exposed to chronic gamma irradiation over a period of years. The site was in a naturally forested zone, about sixty miles east of New York City, and part of the eastern deciduous forest.[10] The land had been farmed and later used as a military base before the laboratory was established. It had been substantially abandoned after the First World War, and a common sequence of natural plant communities had emerged from the abandoned fields. These "successional" communities included shrub and tree stages common on the impoverished, sandy soils of those coastal regions. It had become an oak-pine forest by the early 1960s when the experiment began, more than forty years into the abandonment of the land.

The experiment involved establishing in a plot of Brookhaven's forest a single central source of gamma radiation to provide a gradient of chronic exposure that following the inverse square law, diminished with distance. That gradient offered one set of "controls." The second was offered by a thorough inventory of the stand in advance of the irradiation. The experiment would supply an opportunity to examine how species were affected by continuous irradiation along that gradient over years and determine how the metabolism of the forest was affected as well. This arrangement provided a quantitative measure of the disturbance (the irradiation), data concerning the structure of the community from before the disturbance, and a gradient of exposure from high close to the source to background levels at a great distance.[11]

How would the effects be measured? Obviously there would have to be an elaborate inventory made of the biota before the experiment and at various times subsequently. But there should also be an effort to develop landscape- or community-level criteria of change—a realm in ecological study that was embryonic, if that, in the early 1960s. Even the concept of the metabolism of a forest (total exchanges with the atmosphere through respiration and photosynthesis) was novel and, until then, unmeasured. New equipment was available, and it offered some promise. At that time, we were using the early Beckman infrared gas analyzers, which were capable of measuring carbon dioxide concentrations in air down to the part per million level. That possibility offered an opportunity to appraise both photosynthesis (the absorption of carbon dioxide) and respiration (the emission of carbon dioxide) by the entire forest.

The experiment was established using a ten-thousand-curie cesium 137 (Cs 137) source of gamma radiation suspended on a fifteen-foot tower in the center of the forest. Cesium has a thirty-year half-life (the period in which radioactivity decays by 50 percent), and the exposure rate would be substantially stable over a ten- to twenty-year period. The cesium was in a cylindrical configuration that would allow it to be lowered into a lead cask and thereby shielded so workers could enter the forest safely.[12]

The effects were conspicuous within months. They were not subtle, although ample detail was available in all the studies to provide data on slow shifts in the abundance of species, including any invasive or exotic organisms.

I was astonished as the story wrote itself in terms that were surprisingly familiar and I saw how fundamental the emerging pattern of damage to the forest as a whole was.

There was no reason to think that ionizing radiation had been an effective selective factor in evolution. While we knew from Sparrow's work that there were large differences in vulnerability to the effects of ionizing radiation among species of plants, we had no intimation that we were exposing a new natural law: the ineluctable sequence of disturbance of all sorts on natural plant communities. Yet the gradient established by chronic gamma exposure was easily identified as the gradient of impoverishment observed widely in natural plant communities but rarely described or discussed.[13]

In our Brookhaven experiment, the pines were by far the most sensitive, but after several months' exposure all species of trees disappeared at some point as one approached the radiation source. At higher exposures than those affecting the tree populations, the high shrubs disappeared, and at slightly higher exposures, the low shrubs, then herbaceous plants, then mosses, and finally lichens vanished. Among the lichens, the crustose lichens that cling closely to bark and rocks were by far the most resistant. Here we have the gradient of impoverishment of terrestrial plant communities writ large.

Once defined, the gradient seemed obvious and predictable. But all were surprised at the commonness of the result—a pattern duplicated in the passage from forest to montane tundra, and from the forested zones of the Northern Hemisphere to the Arctic tundra, and thence to the sparse vegetation of the high Arctic.[14] It will be immediately recognized, for example, by those who enjoy the eastern North American mountains: increasing elevation brings changes in the forest, first, from deciduous trees to coniferous, and then to diminished plant size and replacement of the forest by shrubland. At higher elevations with more severe climates, the shrubs are replaced by tundra, and the tundra itself changes with elevation and exposure to mosses, and further along the gradient, to lichens and lichen-clad rocks.

So it is that systematic acute and chronic disturbances produce systematic increments of impoverishment, now classified formally and recognized for what they are, no matter the cause. Oxides of sulfur, for instance, downwind of smelters in Sudbury, Ontario, produced a similar pattern of changes in the boreal forest of that region, according to a well-defined study by Eville Gorham and Alan Gordon.[15] Lengthy chronic disturbance favors smaller-bodied, rapidly reproducing organisms over the larger-bodied, long-lived forms that often dominate landscapes. Dense forests yield to savanna, savannas yield to shrublands and grasslands, and so on down the chain of impoverishment. But the species change as well from those specialized forms of well-integrated, long-standing communities to the wide-ranging hardy generalists of chronically disturbed sites.

At the cellular level, the implications of such chronic disturbances in general are no less systematic and tantalizing. While ionizing radiation is unquestionably a mutagen, many other factors including a wide range of chronic disturbances also produce mutations. It is not surprising that organisms that are hardy in extreme environments are also resistant to a powerfully mutagenic force such as ionizing radiation. Extreme environments, such as the surface of bedrock, subject to rapid changes in extremes of heat under the sun as well as to moisture, are also mutagenic, and it is reasonable to assume that selection over many generations has produced communities in those environments resistant to further mutagenesis. Such communities have long been recognized as marked by an unusually high frequency of polyploid species—that is, species having two or more sets of chromosomes.[16] One of the most important observations of Sparrow and his colleagues at Brookhaven was the greater resistance to ionizing radiation of polyploid strains.

But the change in the structure and composition of a natural community in response to a measured chronic disturbance was by far the most compelling lesson. It set out in vivid terms the key concept of systematic biotic impoverishment as the product of chronic disturbances of many types. It defined impoverishment as the inverse of succession, the repair of natural vegetation after acute disturbance, whether mechanical, chemical, or climatic. That repair process has been seen most explicitly as the "field-to-forest succession" defined for the eastern deciduous forest region of North America.[17] There is a congruent succession from impoverished sites that have been released from chronic disturbance to develop toward the naturally stable or slowly changing vegetation of the region, the "climax" as defined by earlier botanists.[18]

The patterns of changes in biotic structure and potential in response to changes in environment are now predictable in both directions, destructive and developmental. Chronic changes in environment that are global produce increments of biotic impoverishment that are global as well. Nothing escapes: parks, reserves, gardens, farms, and forests—all are vulnerable and respond as the environment departs from the general long-term stability under which the normal natural communities developed. Climatic disruption is such a change. So are all forms of chemical disruption from whatever source, although not necessarily global. As these changes proceed rapidly relative to the generation times of the species, impoverishment is intensified and becomes endemic.[19]

The insights of those early studies of ionizing radiation, a factor that could be controlled and measured, and that struck life at its genetic core, were new and surprising. While they were of wide basic interest, at least to ecologists, they had little effect on the scale of recognized hazards posed by nuclear reactors and no effect at all on the reconsideration of liabilities for damage under the Price-Anderson Act. The generally acknowledged primary hazards from reactor accidents remained direct exposures of people either to external radiation or radioactive particles (radionuclides) absorbed from air, water, or food into human tissues.

These studies over years confirmed a long-held perspective of scientists that if humans are protected from radiation hazards, all the rest of nature is safe. The reason is straightforward enough: lethal or crippling mutations are a serious matter in human populations because we cherish individuals. Lethal mutations are viewed as trivial in other populations where individuals can be lost without serious human concern. Further, there is no clear threshold for deleterious effects in either plants or animals. Even low exposures are widely recognized as potentially mutagenic.[20] The criteria that protect people from exposure, if strictly observed, will also protect all of nature.

The concerns about the safety of nuclear energy and its development for nonmilitary purposes that prompted our Brookhaven research and much more in other research centers, such as Oak Ridge National Laboratory in

Tennessee, were well justified by experience over subsequent years. The US Congress' adoption of the Price-Anderson Act in 1957 was an early recognition that accidents in the development of nuclear power and weapons were inevitable. The near disaster at Three Mile Island nuclear power station in 1979 in central Pennsylvania was the first and, so far, only use of the Price-Anderson Act. The research and reviews that the act supported showed that the accident was in fact substantially contained, despite its potential.[21]

On the global scale, the Three Mile Island accident was followed by the now-well-known 1986 explosion at Chernobyl, which resulted in the deaths of several workers and contamination of extensive areas of Ukraine with long-lived radioactivity.[22] There had also been a serious yet carefully concealed release in a huge explosion in 1957 of a tank storing radioactive wastes from a reprocessing plant at a secret Soviet reactor site, the Mayak nuclear complex near Chelyabinsk, a thousand miles east of Moscow. No official announcement was ever made of the accident. Information available on it was at first pieced together from independent reports over a decade and more by various scientists. In 1980, twenty-three years after the disaster, a Russian dissident, Zhores Medvedev, having written earlier about the disaster, published a book in the United States outlining basic facts long formally hidden.[23] Later, others filled out details of what became known as the Kyshtym-57 disaster. A group at the Oak Ridge National Laboratory showed in 1980 through a comparison of USSR maps over time that numerous villages had disappeared. Estimates of the effects varied widely. According to one report, 270,000 people in Chelyabinsk and the neighboring towns were exposed to "dangerous" levels of radioactivity.[24] The heavily contaminated zone apparently involved three hundred square miles northeast of the site and has been recognized subsequently by visiting scientists as the most heavily contaminated site in the world. Estimates of mortality ranged widely to as many as 30,000 or more over twenty to thirty years.

The disaster at Fukushima in 2011 illustrates dangers of a different sort. In March of that year, an earthquake off the Japanese shore at the edge of two continental plates suddenly shifted the seabed and started a huge wave expanding across the Pacific. The wave, its energy undiminished by distance or any shielding at all, thundered ashore on the nearby island of Honshu, and spread devastation over a thousand square miles of coastal cities, villages, farms, and industrial centers. World attention to the devastation of the earthquake, one of the most violent and protracted ever recorded, and crushing floods from the tsunami, a word invented in Japan because of the frequency of such disruptions, soon gave way to concern about hazards posed by the destruction of three of the six nuclear reactors at Fukushima on the coast north of Tokyo. Damage to their cooling systems made them vulnerable to massive leaks of radiation. Explosions of hydrogen gas, generated by the damage and flooding, contaminated air, water, and land over many square miles with radioactivity at levels that would be seriously hazardous to people. Food, including milk and vegetables produced on farms miles away, was contaminated. Years later, radiation was still leaking from the damaged reactors and full control of all the reactors may not be regained for years to come.

The collapse of the reactors at Fukushima and devastation they spread in Japan struck a nerve throughout industrial society. Events at Fukushima precipitated a sweeping review of all reactor accidents by a physicist long familiar with nuclear energy, Thomas Cochran of the Natural Resources Defense Council (NRDC). Cochran observed that standards for reactor safety set by the US Nuclear Regulatory Commission's Committee on Nuclear Safeguards established a meltdown frequency of less than one in ten thousand reactor years. But worldwide, he showed, the reactor industry is not meeting even this standard by a factor of ten or more.[25] The failures present immediate hazards not only to those who dwell near reactors but also to all those downwind within miles. And the hazard is to all the nearby land, which could be contaminated for at least several decades at levels that exclude human use—a piece of the world, vitally used, suddenly rendered unusable by an immutable and uncompromising human-produced biophysical disruption.

That's what occurred at Fukushima: a densely populated segment of a large Japanese island suddenly became forbidden land. Had a neighbor—North Korea, for instance—usurped the area, the act would have been widely considered and virtually universally justified as *causa belli*. It was a severe blow that caused at least a momentary rethinking of the expansion of nuclear energy in both Japan and other nations including the US. Of the 439 reactors operating in the world in 2015, 99 were in the United States and 58 in France.[26] Each of these in two densely occupied nations has the potential for an accident on the scale of Chernobyl or Fukushima.

In these and the seven other partial reactor meltdowns, the hazards stemmed from direct, acute exposures to humans at the time and the distribution of radioactive wastes downwind. Plant communities were exposed and affected, but those effects diminished over years as normal succession proceeded. The big hazard was long-lived residual radiation from spent fuel debris. External exposures in such circumstances are primarily from gamma emitters such as Cs 137 with its thirty-year half-life. Cesium competes with potassium in all life processes and becomes with other radionuclides, especially strontium 90 (Sr 90), an internal emitter in plants and animals, including people. Sr 90 is similar in behavior to calcium, and settles in bones where its beta emissions become a potential cause of leukemia and other cancers. Iodine 131 (I 131), a gamma emitter, has a short (eight-day) half-life. It is, however, quickly absorbed and accumulated in the mammalian thyroid, where it poses a serious risk, including cancer, to human health. Other isotopes that can be widely distributed in accidents and by bombs include plutonium, which is an alpha emitter and has a long half-life (twenty-three thousand years). Regions contaminated with plutonium present a serious direct hazard to humans from the potential ingestion of plutonium-contaminated air or food as well as the nearly certain potential of the alpha emissions for causing cancer. Such regions— large areas of otherwise-essential landscapes that have already been contaminated, as in Russia, Japan, and those islands of the South Pacific that were in the fallout fields of weapons, will be reoccupied only at serious human risk.

Descriptions of the benefits of nuclear power seldom discuss the problem of management of reactors that have completed their useful lives, have become damaged, or are too expensive to operate. Disassembly is expensive, demanding, and includes the handling of intensively radioactive debris that in effect requires eternal care. The expense of decommissioning a reactor can easily run in a few years to a billion dollars. Such costs have seldom been considered when reactors have been proposed and built. And there is still no long-term solution to the necessity of storing and cooling used fuels rods—a major hazard for all reactors. Robert Alvarez, a physicist and senior policy adviser for the Institute for Policy Studies, wrote in 2012 as follows about the fuels rods already accumulated in the United States:

> More than 30 million highly radioactive spent nuclear fuel rods are submerged in vulnerable storage pools at reactors all over the United States. These pools at 51 sites contain some of the largest concentrations of radioactivity on the planet. Yet, they are stored under unsafe conditions, vulnerable to attacks and natural disasters.[27]

To be an acceptable, viable, and enduring source of energy, the nuclear energy industry would have to become a fully closed system. Despite all the inventiveness and money that governments worldwide have applied to the problem, finding an acceptable system for permanently isolating (protecting the entire biosphere from) high-level radioactivity has proven difficult to impossible and becomes more awkward daily as the human presence expands to the far reaches of the earth. Those difficulties increase the risks presented by the rapidly accumulating spent fuel rods and other highly dangerous wastes as aging reactors are retired. Meanwhile, other technologies offer cheaper and less hazardous sources of energy. Renewable energy can often be produced close to the point of use in small facilities without expensive transmission lines running miles from giant nuclear power plants that require abundant cooling water and relentless attention to safety.

Ironically, the nuclear energy industry is in one important respect a model for the world because of its success in operating, hand in glove with government, a technology that is so relentlessly demanding and hazardous that it requires a system for management that is, with the exceptions noted above, tightly closed from start to finish. The public expects to be protected completely and without fail from exposure to ionizing radiation. Given the inherent dangers of nuclear generation, it has been met by and large with astonishing success. But for its fatal flaws, it is the most outstanding and possibly unique example of an attempt to establish a giant, closed-cycle industry. The protection of common interests in a healthful and sustaining environment requires, as I discuss in a subsequent chapter, that all industries be closed in the sense of containing their wastes, thereby preserving intact the common property resources that they and all the rest of us depend on. In the case of nuclear energy for power, the industry is not completely closed—and cannot be closed. Nor can it be separated from the potential for making nuclear weapons. The leaks, the potential for disastrous accidents, unresolved problems of radioactive wastes, and the proliferation of weapons all work to prevent further development of nuclear energy.

Some interests, nevertheless, still suggest nuclear energy as the potential replacement for fossil fuels, especially appropriate for centralized power plants producing electricity. The distinguished scientist James Hansen and others, acutely aware of the destructive potential of the climatic disruption and the urgency in abandoning fossil fuels that contribute so heavily to greenhouse gases, have recently urged expansion of nuclear power as the only source of energy large enough to offset the closing of coal-fired and other carbon-intensive power plants.[28] I believe that taking this route would be a mistake. First, as we've just seen, nuclear energy must be a closed system, able to account safely at every stage for all its radioactive materials. Yet even now, more than sixty years into the technology, there is limited prospect of systematic, secure long-term storage of highly radioactive wastes. Second, the necessity and appropriateness of retaining centralized power production in large plants is diminishing as renewable energy sources are developed and energy production can be close to the point of use. Third, the reactors themselves are a hazard of a scale increasingly seen by the public as unacceptable.[29] It is, moreover, difficult to separate reactor fuels from bomb production, increasing the risk of the theft of nuclear materials and further proliferation of nuclear weapons. Finally, the safety and security of the reactors as well as the nuclear weapons themselves remain a serious challenge from internal failures and external assaults.

If we consider the record involving bombs, pure instruments of destruction, the story is little short of ludicrous.[30] While the present (2015) stockpiles of weapons are much diminished from their height during the Cold War, thousands remain around the world as other nations struggle, foolishly, to join the nuclear club. Anticipating all potential circumstances under which a weapon might be used in anger is difficult, but reciprocity seems a reasonable assumption if any nation succumbs to such a step. Just as displaying a gun in anger invites another's bullet first, so with nuclear weapons. Use of one such weapon in a world armed with many is, in all likelihood, suicidal. The weapons, despite the folly of many governments and governmental officials, and despite their expense and legions of admirers, have no use beyond the cultivation of terror. The greater hazard appears to be that a terrorist group, immune to a nuclear reprisal, will acquire and use such a weapon. That possibility is a powerful reason for abolishing all weapons and the potential for assembling them anywhere.

We cannot escape the brutally critical perspective of Hannes Alfven, a Swedish Nobel laureate in physics who wrote with penetrating perception that

> fission energy is safe only if a number of critical devices work as they should, if a number of people in key positions follow all of their instructions, if there is no sabotage, no hijacking of transports, if no reactor fuel processing plant or waste repository anywhere in the world is situated in a region of riots or guerilla activity, and no revolution or war—even a "conventional one"—takes place in these re-

gions. The enormous quantities of extremely dangerous material must not get into the hands of ignorant people or desperados. No acts of God can be permitted.[31]

The lessons of nuclear energy are powerful. As they laid themselves out before me over many years, they proved that the dominant worldview of politicians and governmental specialists with respect to nuclear weapons and nuclear energy is warped, if not flatly false, and has been so for decades. Even now after much experience and many sad lessons a revision of perspectives is urgently needed. Nuclear wastes and the radiation exposures they produce are a direct attack on the human genome. They are, however, not unique in that attack. The same or similar dangers lurk for a host of other industries that have established systems for profiting by sharing their wastes with the rest of the world and thereby corrupting the commons. Examples such as the use of coal as a source of energy are abundant. Quite apart from the physical devastation of surface mining including mountaintop removal in West Virginia and open mines in Wyoming are the waste products of combustion released into the air by power plants. Soot, small wind-borne particles of black carbon, vaporized mercury, and oxides of sulfur are all incidental contributions to the atmosphere from coal-burning power plants—progenitors of public morbidity and mortality for industrial profits. Modern industrial agriculture, discussed later in this book, is rich in materials such as herbicides applied in agricultural fields that migrate into water supplies and, carried by air, affect all life elsewhere. Those wastes become a part of common property and join the host of chronic disturbances driving biotic systems around the world into progressive impoverishment. The substances also contaminate human food, accumulate in people as chronic poisons, and produce effects across the spectrum from temporary and mild ones, to sterility, genetic dysfunction, and crippling cancer.[32]

Because the entire nuclear program emerged from governmental initiatives and laboratories as needs were discovered within the nuclear industry, the government's regulatory structure arose more or less continuously and naturally in response to needs. The hazards intrinsic to reactors are enormous, and require relentless scrutiny by operators and inspectors at every level. The wastes, as is now well known, are poisonous, long lived, and often cannot be captured if released. They move unerringly into living systems, where their effects can be devastating. Human lives and large areas of land can slip into oblivion from leakage at any stage. No insurance, not even the Price-Anderson Act's guarantees, cover such losses, which may accumulate over decades as the effects of unmeasured, impalpable, and quite possibly unrecognized exposures emerge.

The nuclear industry can exist only if the public is confident that the entire fuel cycle from mine through manufacture and use in power production and on to waste is tightly controlled—a closed system that contains all its noxious products perfectly isolated, not only from direct human contact, but also from drifting into the interlocking cycles of the biosphere. Such responsibilities require governmental supervision and oversight at every level, from mines to the delivery of the product, whether related to weapons or domestic electric power. As Cochran and his NRDC colleagues have shown, the effort has been astonishingly effective, but far from perfect. And worse, the cycle has not yet been closed, and the possibilities for closing it appear to be diminishing with experience and time. An expensive deep mine in Yucca Mountain in Nevada was planned as a final burial ground for highly radioactive wastes, but it proved vulnerable to potential leakage and has been abandoned.[33]

Quite apart from nuclear weapons, the advent of nuclear energy in general has changed perceptions of the world. Experience has produced surprising lessons concerning the circulation of nutrient elements, toxins, and wastes throughout the biosphere—new insights into the importance of chronic disturbance, including climatic disruption, in causing the systematic impoverishment of natural communities globally. The recognition that the industry must be tightly closed, have no leaks at all, has emphasized that there is no "threshold" below which exposures to ionizing radiation or, more recently, other toxins are safe. Industrial systems that produce toxins quite apart from radioactivity must be closed as well. Suddenly the "commons" has been redefined as the biosphere, the whole world, and the biosphere itself dictates details of industrial design.

The world environment of all life is different now, and a probing review of its potential and management is required if it is to continue long as the habitat of Homo sapiens supported by a hundred million or more of other mutually dependent species.

DDT DRIVES A GEOCHEMICAL
TEMPEST

Although all our yesterdays may have lighted fools the way to dusty death, it is not required that all generations take such an inglorious route.... It is harder to behave foolishly when one reflects on the consequences of one's actions.

—Preston Cloud, *Cosmos Earth and Man*

The sunrise promised another glorious June day in the woods of Maine. Not a leaf fluttered in the early morning air. Nothing moved. No bird. No insect. The forest was silent. No white-throated sparrow called forth identity. No pileated woodpecker's relentless search of enfeebled tree stems for insects echoed. The forest was still. Deadly still.

I stood transfixed for minutes. The boreal forest in all its permutations is rarely silent. It is the epitome of life itself. While it does not hold the richness of the tropics, it is very much alive, full of treats for the alert. It is not silent.

It was 1958, and I had a mission. I was picking up trays I had put out earlier to catch some of the insecticide dichloro-diphenyl-trichloroethane (DDT) that was being sprayed from the air on the forests of both Maine and the province of New Brunswick. The spraying was insurance. It was to protect the trees, especially the balsam fir (*Abies balsamea*) and white spruce (*Picea glauca*), both successional and widely used for pulp, from the ravages of the spruce budworm (*Dionea choristoneura*), an indigenous pest. I wanted to establish just how much DDT was actually reaching the forest, how much reached the ground, and ultimately, how long traces remained on plants and in soils. My techniques were simple. Before spraying began, I had rolled out sheets of aluminum foil of measured size, placed them in critical spots, and collected them immediately after the spraying. The residue was washed from the foil in the laboratory at the University of Maine and the total quantities determined.

DDT was sprayed from the air in an oil solution calculated to spread one-half pound of DDT per acre. The vagaries of wind, aircraft navigation, altitude, and equipment, however, made any quantitative objective in coverage nominal at best. The balsam fir forest I was sampling had been sprayed hours earlier, but the effects of the insecticide over the ensuing twenty-four hours went far beyond reducing the budworm populations.

I was new to the project and impressed with the conspicuously destructive effects of the budworms in stands that had been purposefully left unsprayed. It seemed reasonable at the time that the annual spraying was necessary to prevent devastation of the trees by an otherwise-voracious and highly destructive indigenous pest.

It took some time for a growing cadre of US and Canadian colleagues to realize that, although spraying saved the trees annually and reduced the budworm populations, the small cluster of survivors emerged in the spring into a new crop of rich green leaves and the populations exploded to threaten the forest once again, unless sprayed. It was a perfect system with both the budworm and sprayers enjoying success. Losers were all the other organisms affected: those who had once inhabited the silent forest.

The experience with DDT in the boreal forests of Maine and New Brunswick was a rude awakening. By the 1960s, DDT use had become virtually ubiquitous. When I moved in 1961 from the University of Maine to Brookhaven National Laboratory on Long Island sixty miles east of New York City, DDT was being routinely sprayed from the air to control the salt marsh mosquito. DDT had seemed the obvious cure to a persistent plague, but doubts were emerging. Avian carnivore populations such as the peregrine falcon (*Falco peregrinus*) and osprey (*Pandion haliaetus*), widely separated by habitat, had been declining without obvious cause.

Over the ensuing years my colleagues and I conducted a local survey of marsh and estuarine species, water, and soils that revealed a devastating contamination with DDT. Concentrations of residues varied from a few parts per billion in water to hundreds or thousands of parts per million (ppm) in some fish and birds. In fact the entire food web carried concentrations within one factor of ten of acutely lethal levels. A factor of ten sounds large, but when concentration factors of a hundredfold to thousandfold are common, one factor of ten below acute lethality was as close to lethal as a living specimen could exist. The local osprey population on Fisher's Island was not reproducing and the populations had by 1965 crashed to near extinction. The ospreys' large size and specialized role as top carnivores in the food web made their disappearance significant—a striking symbol of incremental biotic impoverishment. The study we conducted was widely read and remains one of the best examples of the contamination of a food web with a persistent toxin.[1]

The studies in Maine and New Brunswick ultimately showed that DDT residues had a mean residence time in soils of a decade or more.[2] They also brought recognition that aerial applications inevitably resulted in contamination far from both the place of application and the target. It took little imagination to realize that the high-pressure spraying of DDT dissolved in oil inevitably produces small droplets that evaporate, leaving a DDT residue, a small crystal, that can be carried far and wide on air currents before being washed out or precipitated on land or in water. DDT is virtually insoluble in water, where its concentration will be in the range of one or a few parts per billion. DDT is, however, highly soluble in fats. On the basis of solubility alone, plants that contain oils (such as algae) and animals accumulate DDT from contaminated water. There was no mystery, then, in how mountain lakes came to have DDT residues and why almost all animals, including people, by the mid-1960s carried residues in their fatty tissues.[3]

Although it was difficult to determine what the effects of such DDT residues on people might be, it was quite certain that the residues did not improve anyone's health.[4] And it was also certain that low concentrations in air and water were amplified in food webs manyfold, sometimes hundreds of thousands of times. In such circumstances, it is extremely difficult to suggest that there is any level of use that is safe.

Meanwhile, other global changes affecting human health and the biosphere were accumulating as industrial activities gained momentum in the decades following World War II. Agriculture had been expanding as an industrial enterprise for decades following the demands of rising populations. Opportunities for profits soared with new petroleum-driven farm machinery, commercial fertilizers, and pesticides. Industrialized agriculture depended increasingly on simple, inexpensive, broadly effective, persistent chemicals of low human toxicity. From that standpoint, DDT seemed a perfect insecticide and was relied on heavily.[5]

DDT and allied persistent poisons (chlorinated hydrocarbons) were considered safe by many because their toxicity to humans appeared to be low. They were also inexpensive, easily available, and widely distributed in various formulations. They, too, as radioactivity from nuclear bomb tests, simply on the basis of massive dissemination and persistence, became global contaminants of air, land, water, and life to the very depths of the seas.

These two different crises of earth's chemistry, the global contamination with radioactivity and long-lived pesticides, overlapped in time and the details of the hazards they produced. The contamination of virtually all living systems was both a surprise and warning. It was a surprise to discover that poisons used only near the surface of the land could be circulated as widely in air and water as debris from large nuclear weapons injected explosively into the stratosphere. And it was a further unwelcome surprise to discover that life on land and sea, including all humans, carried measurable burdens of these toxins.

From an environmental standpoint the changes constituted a chronic disturbance and contributed increments of biotic impoverishment, as described in chapter 2. From an individual human standpoint, there was a wide spectrum of possibilities from morbidity to mortality. The toxins accumulated in human fat and were most likely to be mobilized during illness, and the symptoms confounded with those of other diseases.

These were powerful lessons, pieced together by scientists over years, but they offered clear evidence of a new realm of threats through channels no individual can avoid. Human activities now affect the chemistry of life globally—all life. Worse, each individual carries a personal burden of exotic chemicals never considered appropriate in food. The number of these exotic chemicals is large, and frighteningly so. No one is above the fray. No one escapes.

If, in the fullness of time and life, governments had explored well in advance the wisdom and interest in making such changes in the chemistry of the biosphere as a whole, there might well have been strong objections. If the "benefits" of nuclear weapons' tests and insect control had been listed ahead of time as the emission of tons of radioactive debris into the atmosphere and a global contamination with poisons harmful to humans and much wildlife, objections among the public would probably have been shrill. These questions were not asked in advance. When they were asked later, it became clear that there were profound issues of human rights involved as well as corporate and governmental prerogatives. Governments, including elected officials, nominally the custodians of human rights and public welfare, had not only allowed the invasions but also had joined actively in causing them and in many instances continue to do so. Once the programs gained momentum, they became commercial, self-sustaining, financially profitable, and approved or at least tolerated by government, difficult to deflect. Recognition of the scale of the intrusions of these programs into biospheric cycles as well as human health and welfare came slowly, delayed by both corporate reluctance to accept the reality of the problem and governmental intransigence. It came as the information gradually filtered from science into the public realm and interest was aroused.

As recognition of the ubiquity and harm associated with these programs spread, despite their justification as immediately necessary, the role of government in a more intensively used world appeared in a new light. The programs raised questions anew about the central purpose of governments, and how well served were the public interest, human safety, welfare, and environmental protection. These questions boiled to the surface in interesting, even exciting ways through the 1960s and 1970s, often with an emphasis on DDT in particular, whose history stands as a warning and powerful model.

* * *

DDT's insecticidal properties were discovered in 1939 by Paul H. Muller, a chemist in Basel, Switzerland, who was screening chemicals for potential use as insecticides. Muller later received the Nobel Prize for his discovery. DDT was used by the US military during the Second World War quite safely and effectively as a powder to control lice. Its virtual insolubility in water meant that it could be used with relative safety around people as a powder. Domestically, around military bases in coastal zones, dissolved in oil, it was sprayed to control salt marsh mosquitoes. It was subsequently released for civilian use in 1945 and quickly became a primary remedy for control of insect pests in agriculture, forestry, and public health. Its chemical stability and persistence assured both effectiveness over long periods and potential for accumulation in troublesome places. Experienced field biologists had warned from the beginning of civilian use of the hazards of indiscriminate distribution.[6] It remained, however, for Rachel Carson, a brilliant biologist and professional writer for the US Fish and Wildlife Service, finally to bring all the shards of evidence together and lay forth the story in lucid detail in the *New Yorker* and subsequently as a book, *Silent Spring*, published in 1962. The book became the most important environmental treatise since *On the Origin of Species* by Charles Darwin and *Man and Nature* by George Perkins Marsh a century earlier.

Silent Spring brought a fusillade from the well-dug-in batteries of industrial interests tied to petrochemicals, agriculture, and public health. Many of the criticisms were unrestrained, slanderous, and baseless. Most were designed to protect the flow of profits. Some emerged from public health specialists who believed alternatives in controlling diseases such as malaria were worse or ineffectual. But topic by topic, increment by increment, Carson's assertions were proven correct even though she had worked with fragmentary data. She brought to the problem a brilliant mind armed with a comprehensive understanding of nature and natural systems as well as a talent for vivid writing, honed over decades.

Carson knew well that living systems have evolved not only to use the climatic and other environmental resources of the earth but also to accumulate chemical substances necessary for their own metabolism. The chemical requirements are not uniform. They vary from place to place as geology varies, and from one community of plants and animals to another depending on evolutionary history. Potassium, calcium, and iodine have special essential roles in life, and it was not surprising that when similar but radioactive elements such as Cs 137, Sr 90, and I 131 were produced in nuclear explosions and contaminated the global atmosphere, they, too, turned out to be concentrated in living systems. The US political and military interests who had managed to persuade a naive Congress that the atmospheric "testing" of nuclear bombs was appropriate denied any contamination beyond a local one until the weapons of the 1953–1958 period had contaminated air, water, land, and life in both hemispheres.

The convenient initial assumption made—or at least promoted—for decades about chemical wastes in air and water has been that the world is large and capable of rendering virtually any contamination safe through dilution and decay. A central principle of toxicology was and remains "dose makes the toxin." Conversely, in this view, reducing the concentration by dilution produces safety. It is a convenient assumption, obviously dangerous in a biosphere specifically equipped by evolution and history to sort, accumulate, and recirculate elements and substances essential for life. Carson smashed all those attractive and convenient shibboleths of industry, commerce, and business-friendly government. Poisons do not necessarily decay into innocuousness. If they decay at all, their decay products may remain toxic indefinitely. DDT residues and petrochemical wastes, for example, persist in soils for decades.[7]

When these issues became clear to the public after publication of *Silent Spring*, many began to realize that the model of the world being advanced by economic interests was inverted, false at the core. Gradually the public interest was redefined in the eye of a growing fraction of the public as the protection of the purity of air, water, land, and the opportunity for life—all common interests and common property. These qualities were a human birthright, not property to be claimed for commercial profit or political advantage. The free-market economic model was eating those very common property resources that governments exist to protect, not to commit to progressive commercial corruption. The juggernaut turns from its course, but slowly, if it turns at all.

I was personally astonished and troubled by the intensity of opposition from nests of supporters of the status quo in the use of DDT and other pesticides. Letters and assertions appeared regularly challenging data and the integrity, purpose, and propriety of scientists who were attempting to deal with the issues rationally. Years later I learned that efforts were made by agricultural interests to persuade Brookhaven National Laboratory to fire me. I heard nothing of it at the time—a tribute to Brookhaven's leadership, which supported its staff and programs loyally and well. I also had the oft-voiced support of colleagues on the staff there, even when I came later to question, on health and environmental grounds, proposals for the use of nuclear explosions in building an isthmian canal and harbors in remote places.

* * *

It is now late in history of the industrial age. DDT and allied toxins, along with radioactivity, have become classic examples whose action and spread provide insight into the much larger issue of industrial chemical contamination, and indicate the necessity for broad governmental protection of human interests.

In the United States unrestricted use of DDT continued until 1972, when William Ruckelshaus, administrator at the time of the recently established Environmental Protection Agency (EPA), banned it. Twenty-seven years had passed since the compound had been released in the nation for general use as an insecticide and fifteen years had gone by since its devastating environmental effects had been rendered conspicuous to all. Ruckelshaus joined the scientific community in the conclusion that safety required zero emissions. His ruling for the EPA substantially established such a standard for the United States—a significant move for the United States stood as a model for the world. Some, however, argued that continued use in small quantities should be allowed in protecting human health against disease-bearing insects such as the malarial mosquitoes, and such an exception was included in the EPA ruling. The World Health Organization continues to argue that DDT is essential for the control of malaria. These contentions are defended on humanitarian grounds—the immediate risks of malaria in affected zones—despite recognition that even small quantities, regularly emitted, persist and accumulate in living systems to high concentrations. And of course the idea of zero tolerance endorsed by the EPA seemed outrageous to many who were thoroughly impressed with the concept that "dose makes the hazard." The insight that small concentrations of persistent substances can be concentrated to dangerous levels by several mechanisms was foreign to many who do not realize that the energy of the whale is solar energy, captured by algae, largely single-celled plants of the sea, and passed up food webs ultimately to appear as those giant animals and propel them on their missions. Nutrients and toxins follow the same routes.

Meanwhile, ignoring the powerful lessons of radioactivity, DDT, other chlorinated pesticides, and the effects of chronic disruption on the structure of nature, commercial interests have continued to expand the use and release of toxins. They have had the support of the US government; all have proceeded with the assumption of safety in any release of pesticides and other substances unless they were proven specifically harmful. On that basis, direct, proven hazards to humans from lead, arsenic, and mercury to asbestos and many other well-known toxins were recognized as dangerous, and are now controlled in whole or in part in the United States and other nations.

DDT had emerged and entered the public realm through a strange and unconventional route involving use in war to combat serious insect-borne diseases. The "problems" in civilian uses proved to be not only the collateral effects, which were real enough, but also the economic and political disruption that the new insights into persistence and toxicity caused. Just as controlling the spruce budworm in Maine and New Brunswick had spawned a new aerial-spraying industry, so, too, in other aspects of forestry, agriculture, and public health, commercial interests had been developed around the use of DDT and allied poisons. While one might have expected the state agricultural universities to have read the signs long in advance and deflected the emphasis on such poisons, the universities were often centers of resistance to any suggestion of a problem. Many of those schools were in fact heavily supported by the industries making and distributing pesticides. The support came in the form of grants for research on pesticide uses and effectiveness, not for studies of safety or collateral effects. The approach of the pesticide industry was direct through control of money for their "research" in universities, but it was also political through pressure on members of Congress and connections in the US Department of Agriculture. Those connections, once understood, gave government and its allied agencies a strong industrial bias that helped explain why the biophysical problems with poisons emerged not from agricultural research but rather through scattered independent channels, ornithologists, and other curious environmental scientists, and why the resistance to regulation was so ferocious and correction so difficult and long delayed.[8]

The DDT story with a conclusion of no threshold for safe use remains a cautionary tale. It should have established a new set of standards in avoiding poisonous excursions in environmental chemistry. When joined with the experience with nuclear weapons, it should have by now fired a blazing effort in government, governmental agencies, and science in general to define and protect the integrity of biospheric chemistry simply to assure human safety as well as the continuity of a wholesome, secure human habitat. Neither government nor the scientific establishment, including the National Academy of Sciences, was prepared for such a challenge. The corporate bias was pervasive, and despite obvious threats to global issues of human safety and security, dominates to this day.[9]

The initiative in such research shifted to the nonprofit world which has found itself often almost alone in organized support for governmental action. On pesticides for instance, recent studies led by NRDC staff and pursued with the EPA, focus on the widely used commercial toxins sold for both plant and insect control.[10] Such substances are now "licensed" for various uses by the EPA, presumably acting in the public interest. President Richard Milhous Nixon established the agency through an executive order in 1970 partly as a result of the devastating experience with DDT. Congress approved the order, and the EPA became an essential arm of government. The EPA's emergence by that route was a quite-remarkable and important advance in environmental protection. But continued protection requires continuous reinforcement, for the EPA as a public agency must listen to all, including corporate interests.

After the EPA placed restrictions on DDT, commercial alternatives were sought for insect control. Attention focused early on nerve toxins derived from nicotine, the "nicotinoids," now by far the most heavily used agricultural insecticides in the world.[11] They share with DDT the advantages of persistence and broad toxicity. Some of the great variety produced added the further advantage of incorporation into plant tissues, making the entire plant toxic. Little sophistication in ecology was required to anticipate collateral effects, entirely predictable when broadly toxic compounds with long residence times in nature are widely distributed. Yet the EPA licensed these compounds for use in agriculture. The experience with DDT was an inconvenient model, simply ignored. The EPA must listen to all petitioners, commercial and agricultural supporters as well as independent scientific or other objectors, and is unable as a practical matter simply to ban a commercial technique or substance without specific experience. As an epidemic of bee colony collapses struck the agricultural community over recent years, research implicated neonicotinoids. Such connections are always difficult to prove. Proof is demanded, and often contested, by those who benefit from the sale or extended use of the poison. But data are coming forth now as bees in damaged colonies are discovered carrying the toxins. The evidence is powerful that the neonicotinoids are responsible for the catastrophic decline in bee populations. We appear to have another, entirely predictable and unnecessary yet thoroughly devastating crisis in agriculture propagated by a profitable commerce in poisons. And this says nothing about the even more difficult challenges in determining both human exposures and effects. Wide distribution of nerve toxins in segments of the human food web can be assumed to be contaminating humans, and it is no surprise that serious concerns exist over possible neurological effects.[12]

The problem of broad toxicity in contemporary agriculture is not limited to neonicotinoids. One popular aspect of the genetic modification of crops has been the development of herbicide-resistant strains. Such crops can be raised without annual tillage simply by drilling the seeds of the herbicide-resistant crop into unplowed ground and subsequently spraying the resistant crop with herbicide to eliminate competition from weeds. "No-till" agriculture is understandably popular among farmers, but it results in the virtually ubiquitous contamination of groundwater and runoff with herbicides,[13] as well as selective pressure that favors resistance to the herbicides in weed species.[14]

All the difficulties inherent in proving the collateral damage of DDT are present in establishing the hazards of the nicotinoids and various herbicides now in general use. Just as DDT's broad array of uses could not stand close scrutiny, so the general distribution of neurotoxins and plant growth inhibitors are vulnerable to the critical reviews that have shown catastrophic effects, such as the destruction of essential pollinators of crops and contamination of the human food web. Any delay in outlawing the use of those compounds improves their profitability in the market and assures that users will bring arguments aimed at further delay. We have, apparently, learned nothing from experience. The governmentally supported agricultural establishment of the state university system is again busy advancing the interests of industrial agriculture over those of human welfare. The public interest is again left for advancement not by representatives in Congress, where industrial money dominates, but instead by petitioners from outside the system required to produce new data specific to each chemical intrusion. Science and conservation can never expand enough to defend the public interest against such systematic corruption.

The lesson from DDT and radionuclides is clear: human safety and the integrity of structure and function of the biosphere require absolute protection, zero release of any toxin that can accumulate (as opposed to immediate decay),or whose decay products are toxic and can accumulate in any segment of the biosphere. That lesson does not require demonstration repetitively. And responsibility for recognizing that fact lies not with a few independent scientists or specialized nonprofit agencies petitioning governmental agencies such as the EPA but rather with the governmental agencies themselves and the political system that supports them, all of whom have the capacity to learn and predict on the basis of experience and insights. Building such knowledge is the core of science, and accumulating and applying it is a key responsibility of government. The United States has not yet reached that degree of maturity in government or in science. The issue is now large and growing more urgent and critical.

There are, unfortunately, abundant examples of unchallenged environmental corruption. Many thousands of industrial releases, agricultural poisons, and petrochemical wastes, including carbon dioxide and methane as well as a wide range of chemicals used in the plastics industry, are not controlled. For many agricultural fertilizers and poisons, such as herbicides, insecticides, and growth hormones, there is clear evidence of wide circulation, potential migration into human food, accumulation in human bodies, and potential influence on natural communities of plants and animals. The extent of such intrusions is difficult to exaggerate for there are literally thousands of exotic chemicals released with little acknowledgment of the range of effects. Effects are sufficiently diffuse and remote in time and place to be considered uncertain and occasionally potentially immeasurable. It is this asserted uncertainty and nominal ignorance that are drawn on for a rationale for continuing profitable uses despite extensive experience with analogous toxins. The responsibilities for control here have been inverted from restrictive and protective of the human birthright and integrity of global chemistry to permissive destruction by commerce for profit.

The proliferation of releases remains rife with economically attractive false assumptions. The core assumption remains that any "waste" or other chemical intrusion is safe unless proven otherwise, usually as a direct toxin for humans. Even within that coarse limit the scientific enterprise will never be able to meet the challenges of testing thousands of substances used, distributed, and released in one form or another by the world's industries. The current criteria of safety furthermore ignore any influence of the substance on the structure or function of biotic communities. Such effects occur at concentrations that are far lower—in many instances, orders of magnitude lower—than those commonly recognized as toxic to people. The failure to follow commonsense restrictions in releasing industrial wastes and the use of poisons for control of pests is having profound effects. Industrial wastes dumped into the atmosphere travel worldwide. Ultimately residues aside from DDT accumulate in the oceans, washed out directly in rain, or precipitated onto land and washed into streams that reach the coast.

Even more troublesome are substances that have a vapor phase at normal temperatures such as mercury, widely used in mining gold and a product of various industrial processes as simple as burning coal. The vapor, even at low concentrations, is mixed into the atmosphere, spread widely, and condensed onto cold surfaces wherever they occur. Low concentrations of vapor over time can move large quantities of such chemicals over great distances. Mercury, vaporized in lower latitudes in the industrial world, is condensed in cooler Arctic regions, where methylated microbiotically, it becomes available to contaminate mammalian and other food webs. Mercury is only one of the noxious pollutants that follow that atmospheric route to colder regions, where food webs are no less vulnerable than at lower latitudes. The effects can be devastating. Polar bears are accumulators of the mercury used industrially around the world.[15] So, too, humans who consume fish and mammals from high latitudes are exposed.

Significant changes in the chemistry of the biosphere are often dismissed by the wishfully optimistic assumption that human health and welfare can be protected in a contaminated biosphere. The swift and sweeping infusion of radioactivity into every organism on earth as well as the parallel contamination with DDT and other persistent chlorinated hydrocarbons and their effects belie that hope. The noose is tightening year by year as additional research shows how seriously vulnerable humans are to protracted low exposures to common municipal air pollution. One channel is the human endocrine system, which is vulnerable to disruption by small exposures to a wide range of pollutants common in the industrial world.[16] Even more distressing is the evidence accumulating that in utero exposures that occur through mothers breathing the common air of modern cities cause subnormal development of the embryo and the individual as a youth.[17]

The change in the basic chemistry of the environment that the expanding industrial system has been imposing on the world is one of the conspicuous chronic disturbances contributing to the march of biotic impoverishment. But the structure of biotic communities usually moves slowly enough that years of careful measurements may be required to offer compelling data on the changes and their effects. Such a study of effects, however, has been completed recently testing the effects of the increase in nitrogen, usually as nitrate, in the runoff water feeding a salt marsh on Plum Island off the New England shore. The increase in nitrogen has several sources including nitrogen fixed by the high temperatures in all internal combustion engines (especially diesel engines in cars and trucks), runoff from agricultural and lawn fertilizers, and drainage from septic sewage systems. In the marsh the nitrogen is a fertilizer, and as in gardens and lawns, favors the growth of vigorous plant tops at the expense of roots. The fertilization of the marsh, continued over ten years with an artificial supplement to speed the process, produced drastic changes in the marsh quite consistent with that experience: the plant tops were stimulated in growth at the expense of the root systems. Yet it is the roots that hold the marsh in place against the tidal flows. As the root systems began to fail, the sediments in the marsh were washed away, and the marsh began to resemble not a marsh but rather a series of ponds.[18] Such changes over a decade or more are not necessarily noticed at first, but they can be profound in their effects on the landscape and wildlife including spawning fish that normally use the marsh at critical points in their lives.

Despite broad experience with the range of toxic substances produced by the current fossil-fuel-powered industrial society, the US government has been led away from the strict control that scientists might advocate. The European Union, in contrast, regulates on the assumption that wastes are hazardous unless proven safe.[19] Instead of close control, Americans assume a loosely permissive stance that favors industrial release of frequently untested chemicals into the public realm. The process in effect transfers a liability from industry to the public without a charge to industry, to the detriment of all. Tacit acceptance of the liability by the public is, as with the Price-Anderson Act, granting a public subsidy to industry. Without that subsidy, industry would incur a potentially burdensome expense. Instead, industry gains a well-focused financial advantage. The public shares the cost in decrements of health and finance, and in struggles to restore water supplies, the quality of air, and the purity of food.

* * *

We live now in a world in which the global environment is not simply at risk; it is eroding rapidly. And the erosion is not only in the accumulation of industrial scars earlier identified as "sacrificial zones and communities," the mined-out lands deplored by Hedges and Sacco.[20] It is in the far more comprehensive global changes in climate and environmental chemistry now sweeping the earth. The issues are exploding before us from local disasters such as coal mining by mountaintop removal to near-imponderable global crises. We need not look far for powerful examples.

The world was stunned a few years ago when the Gulf of Mexico fell victim to the largest oil disaster that the world had ever known. Rimmed by a thousand miles of marshes, mangroves, beaches, and deeply cut estuaries, including the Mississippi Delta, the Gulf has shared its biotic riches with humans for ten thousand years and more. On April 20, 2010, BP (formerly known as British Petroleum), one of the world's largest corporations, lost control of an oil well in deep waters of the Gulf far from shore. The blowout killed 11 of the 126 workers on the giant platform, and over the ensuing three months, an estimated 4.9 million barrels of crude oil were let loose into the Gulf. Some of that oil was dispersed into the water column by detergents sprayed for the purpose. While the oil so treated disappeared from view, its noxious effects were spread even more widely. Just how serious those diffuse effects are remains an awkward topic for science, but the dispersed oil constitutes a major change in the chemical environment of all life in the region and is vulnerable to all the various concentrating mechanisms characteristic of natural communities. Heavier oil many months later still coated the bottom over unknown areas of the mile-deep water. Birds, fish, and marine mammals were killed over months. Coastal marshes were oiled and may remain so for decades.[21] The dispersed oil, mixed by wind and currents of the Gulf into large volumes of water, moved with Gulf waters through the Straits of Florida into the larger circulation of the North Atlantic Gyre.

BP is but one of several corporate exploiters of oil and gas in the Gulf of Mexico. But the disastrous spill was BP's own, a product of mismanagement, a devastating "error" that affected the entire Gulf. No one had anticipated such an event, although once it was under way it became clear that one corporation had suddenly affected—and by affecting, claimed unto itself responsibility for—the entire Gulf. In the sense of Hedges and Sacco, the Gulf had been made another corporate wasteland. Tens of thousands of residents along the coast have made their livings for generations by harvesting the rich life of the Gulf. Much of that life, if present at all, is now contaminated. One giant corporation, nominally regulated in the public interest by government, nominally responsible and experienced enough to drill in deep water safely, took shortcuts, and almost predictably, lost its gamble. The flood of oil impoverished for the indefinite future a source of sustenance for current residents of two nations on a coastline extending from the tip of Florida to the Yucatan Peninsula. By whose right, we might ask, does such contamination exist, and to whose advantage?

Ecologists, biologists sensitive to the fact that we live in an environment defined by life itself, find events like the BP oil spill particularly terrifying. Such events challenge in increments the very existence of life on earth, starting with the structure of this civilization. The human intrusion, aided by advanced and extraordinarily powerful technology, is now large enough that one single corporate unit, one oil well in the Gulf of Mexico, mismanaged almost beyond belief, can run out of control for months and poison the entire Gulf with a flood of oil. And what are the effects? How long will they last? The questions came too late. They should have been asked—and answered definitively—before the license to drill was issued. Answered honestly, the Gulf would have been seen for what it is: one of the world's richest biotic realms connected tightly to the North Atlantic circulation by the Gulf Stream. Its chemical and biotic integrity would have been seen as essential to the normal function of the marine and coastal ecosystems of the southern and eastern seaboards of North America. Putting those resources at risk of destruction or impairment should not have been a possibility. No chance at all. Now, the cost of the disaster is beyond calculation, and the time over which the cost will be paid is measured in human generations. A monstrous subsidy has been granted to the corporation, now being allowed again to drill in the Gulf, with other oil companies, all reaching for the same oil far below the surface, far from land, competitively, for profits, subsidized by a governmental gift of the entire Gulf of Mexico, from north to south, from the Yucatan to Florida and beyond, from top to bottom and below to wherever the oil is, and the North Atlantic Gyre and the rest of the oceans, for they are all mixed over time. And what is the purpose? The same as other petroleum interests: to acquire, refine, and sell for profit to the public oil accumulated in the global crust over hundreds of millions of years, now to be released over a few years as carbon dioxide and methane to poison an atmosphere as well as cause an open-ended global climatic disruption, a cascade of biotic impoverishment with the potential for destroying this civilization within the lives of people now living.

Public subsidies are gifts from the public to corporations to enable the financial success of corporations or other businesses thought to be performing a public service. No subsidy is larger than the de facto grant being made to the fossil fuel industries in allowing their wastes, especially their waste gases, to poison the atmosphere and change the whole earth. That subsidy is not calculated into the economics of the industry, nor is the commitment of land and water to the mining of the fuels recognized as a part of the subsidy. These public commitments are made through governments. In the United States the gift is amplified for the oil and gas industries, for example, by specific financial subsidies that allow credit to the corporations for the "depletion of reserves," which are in fact public property initially and are opened to industry for mining at modest charge through governmental auctions. Why their depletion should continue to be encouraged with a subsidy is not at all clear except that the corporations exert great political influence. The reserves are "depreciated" in the United States as equipment is depreciated over time, up to an amount equivalent to 15 percent of sales.[22]

But by far the largest subsidy accrues through allowing the release of industrial wastes—greenhouse gases—to accumulate in the atmosphere and undermine the entire human undertaking—our civilization—and the earthly habitat of all life. And this is all for the profits of several large corporations that appear to have no intention of modifying their highly profitable businesses. David Hawkins, writing for the NRDC's *Switchboard* reflected on the corporate view:

> Running a major oil company means not having to say you're sorry. Or at least that's what ExxonMobil's managers seem to believe. Earlier this week the oil giant posted a report on how it plans to reconcile the need to sharply limit fossil fuel emissions to protect the climate with its continuing gargantuan investments in additional oil and gas reserves. Its answer? We aren't changing course. We're going to keep drilling and spilling (carbon pollution at least); the world will just have to live with it.[23]

Is there a compromise to be struck between industrial economic interests and biophysical welfare, including human health and opportunity, now and for the future? Is it even realistic, much less right, to offer industry an environmental subsidy that guarantees systematic impoverishment of living systems globally? Any compromise entails a chronic disruption, exacerbating chronic effects that are cumulative, a constant abrasion that makes "life" more difficult for all. The tragedy of the commons is a universal tragedy from which escape is difficult, once it is entrained.

* * *

No environmental topics have been better studied and tested by contemporary scientific, political, and economic communities than nuclear energy's radioactivity and other hazards, and the persistent pesticides with DDT as the model. Various tribunals and branches of government have also examined these contrasting topics over decades, and cogent answers to questions of management in the public interest have come forth clearly.

For ionizing radiation, the lesson is simple and unequivocal: exposure at any intensity makes a contribution to corrupting the human genome, the entire genetic complement of each individual starting with the zygote at the beginning of each life. Avoiding the panoply of effects is straightforward: ionizing radiation beyond the inescapable naturally occurring background radiation must be contained, completely contained, and not distributed in the general environment at all. There is no safe exposure. That is the conclusion of extensive research over a century and a half. Governmental policies, laws, and rules support it, allowing only limited exposure for workers where low exposures cannot be avoided and medical purposes where there is a clear, immediate benefit to the individual. But the dangers are real.[24] The nuclear industry is ideally a tightly closed system, with no environmental release acceptable to the public. That, at least, is the expectation on the basis of experience. Accidents such as the Fukushima disaster only tighten the conclusion.

With respect to DDT and other toxins, the experience is similarly detailed, but the conclusions have emerged more slowly and less definitively. The economic implications are profound and vigorously defended by large, politically powerful industries. From the perspective of human health and welfare, two principles emerge and are gaining wide support. First, there is no remaining basis for the attractively misleading concept that dose makes the toxin and its corollary, "dilution cures all." These ancient shibboleths of toxicology have fallen before the weight of the heavy hammer of evidence on the endocrine control system, which is sensitive to even minor exposures to a range of chemicals, especially in the developmental years but also throughout life. Again, establishing a safe level of exposure is virtually impossible. There is no threshold. The tolerance level is zero.

No less telling is the evidence from DDT, other chlorinated hydrocarbons, mercury, other metals, and hosts of other noxious products that these substances emitted in low quantities can be accumulated by various biophysical processes to high concentrations in unexpected places. Again, there is no threshold below which use is safe. Correcting the present trends requires that the insights we have now become simple rules that are the sine qua non of civilization from here on. The core concept is essential to human welfare in the narrowest, most tightly focused self-indulgent context as well as the largest sense of preserving the rights of generations to come. It is that the chemical integrity of the biosphere must be recognized and preserved. Again, we return to a central principle described in chapter 2: developing and requiring closed systems for all industries, and ultimately, of course, municipalities and domestic affairs.

In addition to that essential innovation is recognition that the world can no longer offer industries the opportunity of turning segments of the biosphere into sacrificial zones. The Gulf's biotic integrity—that of its waters, estuaries, marshes, depths, and flows, ever at hazard from intensive harvests—is an essential element of the biosphere, a piece of its functional anatomy, integral to the land and sea of the Western Hemisphere. There is no circumstance in which human welfare can be advanced by a sacrifice of the Gulf to corporate purpose. The fossil fuel industry may be large and powerful, but its business success is a double-edged sword. It profits with one swing of the sword, which severs the life support of millions dependent on the biotic integrity of an unpolluted Gulf along with its affiliated dependencies of land, sea, and air. In recovering for another swing, it sells the oil and poisons the planet with its gaseous wastes while collecting public subsidies for the privilege. It is hard to envision such corporate activities in devouring common property at public expense as anything but crude, rapacious, and ultimately suicidal.

We can do better.

Quite unexpectedly the top blew off.

—Archibald MacLeish

The line in the aerial photograph was sharp. It was the frontier, the edge of the industrial world. On one side, the intact moist forest of the Amazon basin; on the other, the soybean flatland that has replaced the forest over a thousand square miles and more. By comparison with one of the richest forests in the world, the soy farm was a desert.

The *New York Times* does not regularly treat the world to such a pictorial feast, a powerful statement of ecology on the front page. Yet there it was in September 2009: a photographic definition of the major transition of our time—the replacement of a self-sustaining biosphere by a fossil-fuel-powered industrial landscape, a piece of a large industrial machine. It was as close to a single figure as we could hope for in depicting the cause and consequence of the climatic disaster that is to become part of an indelible record in the sediments of our time on earth.

It offered a prediction of what is to come. In that context, it was terrifying. In another, the photograph was richly educational and essential news.

The forest, alive with perhaps five different layers of leaves, absorbs the energy of the heavy tropical sun. Most of that energy goes into the evaporation of water. A visitor from the temperate zones has to marvel that the shade and evaporation keep the interior of the forest cool and comfortable, the sun, filtered and remote.

In a brilliant series of studies in the 1990s under the auspices of the Woods Hole Research Center, Daniel Nepstad and his colleagues showed that the roots of those trees penetrate soils to depths of forty to sixty feet and draw water from the entire depth—an astonishing discovery that changed the way scientists think about the largest area of tropical forest in the world.[1] The energy absorbed in the evaporation of water from the leaves of the still-enormous extent of the Amazonian forest is carried as the energy of vaporization into the atmosphere to drive the climatic system of the Northern Hemisphere. The depth of the roots assures the availability of water, even in dry times. Some of the vapor is condensed locally in rain showers downwind of the forest.[2] Some enters the larger hemispheric circulation and is carried to higher latitudes, where cooled, it condenses as precipitation. The abundance of moisture and solar energy (the sun is high in the sky throughout the year in the tropics) means that on a warmer earth, more water will be evaporated, cooling the air and preventing great changes in the temperature of those low-latitude regions.

It is the energy of vaporization that drives the winds and makes the storms and whose transformations are involved ultimately in determining the places and amounts of precipitation. With more energy available on a warmer earth and the consequent increase in evaporation in the moist tropics, the climatic system moves more rapidly and the normal cycles are intensified. Climatic zones migrate, continental centers are warmed, and arid zones expand simply because the configuration of the Northern Hemisphere continents in those latitudes exposes more land area to the migrating aridity. Continental margins, normally moist, become wetter. Floods are more common. And the heat of vaporization, released in the higher latitudes, warms those normally cooler regions differentially—two or more times as rapidly as the mean temperature of the earth as a whole. So the boreal forest zone, normally moist and cool, becomes warmer and, in some quarters, wetter; in others, because it is warmer, drier. Such changes are not simply one-time changes to a new regime. They are a change from substantially stable conditions over centuries to continuous, year-by-year changes in the annual climatic cycle including temperature regimes and precipitation. These are chronic disturbances, a continuous change away from the climatic systems into which the existing biotic systems have evolved and to which they are adapted. And the effect follows the pattern of impoverishment defined in the Brookhaven experiment described in chapter 2 and observed around the world.

The soybean field in the *New York Times* photograph, in contrast to the forest it displaced, is shallow rooted, bright in color, reflective of the sun, hot, and by comparison with the forest, dry. Walking from the forest to the bean field is stunning, not only by the contrast in passing from shaded to bright, but also from cool to exposed, with the sun suddenly heavy whereas in the forest it had been remote and subdued. The implications for the water regime are no less profound, for the amount of water vapor entering the atmosphere from the agricultural field is by contrast with the forest, trifling, and without rain or irrigation, soon exhausted. Precipitation in the Amazon basin as a whole is vulnerable to such changes, and if the transition to agriculture is extensive enough, will be hugely affected. And of course local runoff will also be affected, as will be the soil structure and storage of water. The basin suddenly goes from moist forests to savanna and incipient desert.

The transformation shown so vividly on the front page of the *Times* also entails another essential aspect of the global crisis of climate: the transformation of forest to agriculture results in the combustion, or decay, of the wood and much of the organic matter in those deep soils. The decay becomes conspicuous a few years following deforestation by fires or the harvest of trees as giant termite nests appear above ground, signifying the accelerated decay of the masses of tree roots that add to the billion tons of carbon released annually from the destruction of forests globally. The products of combustion and decay are the same: carbon dioxide, heat, and water.

To most observers, that photograph undoubtedly passed as simply another remarkable depiction of the advance of civilization into the vast forests of the Amazon basin, the continuing triumph of industrial agriculture and inexpensive food over a forested wilderness. To others, however, it is a powerful illustration of the biophysical cost of expanding agriculture into the tropical forests globally to accommodate a surging human population. The transition from forest to single-purpose food production (and associated profits) may seem wise at the moment, but it comes with some large changes in the earth that ultimately threaten food production, forests, and human welfare worldwide as well as the industrial agricultural economy. Where is the wisdom in destroying a thousand square miles of a forest that taps and maintains soils to depths of sixty feet, and extends several layers of leaves to heights of a hundred feet and more, supports thousands of species, and contributes to maintaining a predictable global climatic system—the whole capable of sustaining itself indefinitely and contributing to a global biospheric climatic system no less predictably stable? All this destruction is for the single purpose of producing a crop of commercial value as human food for those who can enter that market and buy food. Such an exchange is surely a short-term gain for a few, if not an immediate disaster.

* * *

For more than eight hundred thousand years before 1880—fully half the period that modern humans have existed—the carbon dioxide concentration of the earth's atmosphere did not exceed 280 ppm. Now, in the first decades of the third millennium, it is rising above 400 ppm—an increase of nearly 40 percent. Unless the rise is checked soon by the combined action of the world's 193 nations, it seems certain to soar toward a doubling of the 1880 levels in the next decades. While the earth has warmed over the last century by an average of less than three-quarters of a degree Celsius (a little more than one degree Fahrenheit), the warming in middle and higher latitudes is two or more times the global mean. Substantial climate-related changes in land and water continue to intensify—and disrupt economic and political systems globally. These trends have been the topic of compelling reports and analyses over more than forty years. The most recent reports include a new generation of scientists speaking clearly for the far-larger scientific community, which is overwhelmingly persuaded of the seriousness of the threats.[3]

Among the public the issue is seen as complex, mystifying many and deliberately confounded by others. The issues are straightforward. The heat-trapping gases are a small fraction of an atmosphere that is roughly 80 percent nitrogen and 20 percent oxygen. Neither nitrogen nor oxygen absorbs energy in the infrared segment of the spectrum of radiant energy received continuously from the sun. Carbon dioxide's share, a mere 0.04 percent by volume, seems trifling. But its capacity for absorbing radiant energy gives it great power in determining the temperature of the earth. And because the concentration of carbon dioxide is low, small changes caused even by annual shifts in the metabolism of a forest or other large plant communities become important.

The temperature of the earth is determined by the equilibrium established between the solar energy absorbed and energy reradiated into the blackness of space. An increase in the heat-trapping capacity of the atmosphere increases the equilibrium temperature. Attention has rightly been focused on the enormous role that the use of fossil fuels and massive defor-estation globally have had, and continue to have, on the carbon diox-ide content of the atmosphere. Carbon dioxide is not, however, the only heat-trapping gas added to the atmosphere by this expanding fossil-fueled industrial civilization. Methane, nitrous oxide, and water vapor are also heat-trapping gases, and each has a role.

Methane and nitrous oxide are both also normal atmospheric trace gases, and although their concentrations are much lower than carbon diox-ide's, both are rising as well. Methane, on a molecule-by-molecule basis, is far more efficient at absorbing radiant energy than carbon dioxide (about twenty-three times as much), and even though its concentration and resi-dence time in the atmosphere are less, it remains as methane a significant contributor to the climatic disruption. It is oxidized ultimately to carbon dioxide, but has a mean residence time in the atmosphere of approximately ten years.[4] Its main sources are fossil fuels (gas) and microbial metabolism in the absence of oxygen (anaerobic decay). Anaerobic decay is common in soils and virtually all wetlands. The digestive processes of ruminants and, to a lesser extent, humans also produce methane. Nitrous oxide is a normal product of the global nitrogen cycle. Its concentrations have risen abruptly as a result of the use of massive quantities of nitrogen fertilizer in agricul-ture. Internal combustion engines with their high temperatures and pres-sures also generate oxides of nitrogen. Water vapor, too, is important as a heat-trapping gas, but the big factors affecting the temperature of the earth are the changes in the long-term concentrations of the other gases. The pri-mary concern at present is carbon dioxide, which is responsible for about 50 percent of the climatic disruption. Methane, as we shall see, remains a serious additional threat, especially if global warming is allowed to proceed on its present course.

The changes in climate already experienced are rocking the foundations of every nation and beginning to destroy some. One of the major effects of increasing the temperature of the earth has long been recognized as an increase in the aridity of continental centers. A glance at maps showing the distribution of natural vegetation is enough to illustrate that a general warming will cause the expansion of arid zones in the Northern Hemi-sphere into natural grasslands now in intensive agriculture. Mexico and the southern and southwestern United States are already affected. Central China is also dry and becoming more so. Sub-Saharan African nations are

in political chaos as drought intensifies, lakes disappear, and the potential of the landscape for supporting agriculture diminishes even as human
numbers soar. Lake Chad is drying up and the potential of the vast Sahel to
support even marginal agriculture is eroding. Progressive biotic impoverishment is a major contributor to the collapse of Somalia into civil war. The
pressures are being felt by the European nations as refugees attempt to cross
the southern European frontier seeking a new life in a habitable zone. Med-
iterranean storms regularly overwhelm boats carrying hopeful migrants
toward Italian shores.[5] South Africa is similarly afflicted, overwhelmed with
migrants from the north, filling the slums of Cape Town and Johannesburg
and seeking nonexistent jobs, all looking for a better place to live in a still-
viable nation.[6]

Although the issue of climatic disruption and its consequences has been
written, discussed widely and available for anyone who cared to notice for
the last four decades, and conspicuously in the public political realm for
more than two decades, little remedial action has been taken. The world
seems paralyzed, unable so far to deflect a human disaster without precedent.

* * *

The potential for climatic disaster was anticipated in the late nineteenth
century, but it was not until the late 1950s that sufficient interest emerged
in the scientific community for Roger Revelle, then director of the Scripps
Institution of Oceanography in La Jolla, California, to take serious action to
explore the details of changes in the atmosphere. He wisely hired a young
oceanographic chemist, Charles David Keeling, to improve measurements
of the carbon dioxide concentration in air. The methods at that time for
measuring such concentrations were chemical, awkward, and less than pre-
cise. They were quickly supplanted by the invention of the infrared gas
analyzer, which used the infrared absorptive capacity of carbon dioxide
to detect its presence. Quantitative detection and measurements virtually
down to the molecular level became almost instantly available. In 1958,
Keeling began accumulating data from air samples from the Scripps Pier
in La Jolla, Mauna Loa's ten-thousand-foot peak in the Hawaiian Islands,
and from a station at the South Pole. The now well-known data that Keeling produced showed an upward trend in carbon dioxide concentration,
and in the ensuing years triggered a storm of concern over earthly climates
and human welfare. But as late as the early 1970s, climatologists and other

scientists, fundamentally conservative and always skeptical, were reluctant to talk about a climatic disruption in the absence of clear data showing an effect on the earth's temperature. There were no such data then available, not surprisingly, for it was not a simple matter to take the temperature of the earth. And there appeared to be no way of looking back thsrough time to appraise the composition of the atmosphere and temperature changes over even the recent past. Today, fifty years later, I own a coffee table made from a cross section of a three-hundred-year-old Douglas fir from Oregon that I acquired in a futile effort to develop an isotopic or other technique for measuring carbon concentrations in air of the past.

That issue languished for several years until an imaginative group of geochemists led by Hans Oeschger of the University of Bern in Switzerland saw the potential for measuring the composition of air trapped in glacial ice at the time the ice was formed.[7] It was a brilliant stroke, stimulated in part by the recognition that glacial ice melting in water hisses softly as air bubbles break free under water. Those bubbles had to be small samples of air trapped at the time the ice was formed. Oeschger and his colleagues cleverly produced records of both carbon dioxide concentrations and, using oxygen isotopes, temperature that ultimately extended to eight hundred thousand years. The data showed not only that the carbon dioxide content of the atmosphere had not exceeded approximately 280 ppm over that period; it also demonstrated that the temperature of the earth marched with carbon dioxide concentrations. During glacial periods, the atmospheric concentration was as much as 50 percent below the long-term maximum.

For the ecologists at Brookhaven in the early 1960s then engaged in studies of the responses of a forest to chronic disruption, the new data from Keeling's research and, later, Oeschger and his colleagues were astonishing. To be sure, there was a systematic upward trend in carbon dioxide of about 1.5 ppm per year observed from the beginning of Keeling's work in 1958. But what caught our attention in particular was the annual oscillation in the carbon dioxide content of the atmosphere apparent in the records Keeling was accumulating. The oscillation was especially conspicuous in the data for the Northern Hemisphere taken from near the summit of Mauna Loa in the mid-Pacific, where the air was presumably well mixed. The atmospheric carbon dioxide concentrations reached a maximum in the late northern winter, and dropped through the northern spring and summer to a low in September or early October, and then rose again through the winter to a high in April. The changes were hemispheric and far in excess of any conspicuous human influence.

My colleagues and I were then attempting to measure the metabolism of forests using carbon dioxide exchanges to measure carbon uptake from the atmosphere (photosynthesis) and carbon releases (respiration). In the course of this work, we were also accumulating atmospheric data from central Long Island. These data, too, showed the annual oscillations evident in Keeling's data, but with even greater amplitude. A reasonable assumption seemed to be that the oscillation reflected the seasonal metabolism of the plant communities, especially forests because of their stature and areal extent. The predominance of forests in the Northern Hemisphere, particularly at the higher latitudes where the amplitude of the oscillation was higher, pointed to their seasonal metabolism as the cause. That conclusion seemed to be confirmed by other observations. The oscillation, for instance, was reversed in the Southern Hemisphere and conformed to the pattern of the southern seasons. Further, simple estimates based on our other studies of forest mass and structure showed that forests and their soils globally contain a pool of carbon large enough, if released, to affect the atmosphere significantly.

If we were correct in these assumptions, the metabolism of the hemispheric terrestrial vegetation—especially in the Northern Hemisphere, where there is more land—was a crucial factor in changing the composition of the atmosphere by ten to twenty ppm twice annually in the mid-latitudes. Such a change involves several billion tons of carbon dioxide. There was no question in our minds: any change in the metabolism, a shift in the seasonal storage or release of carbon by plants, particularly forests, had the potential for affecting the earth as a whole in a short time. Suddenly, the photosynthesis and respiration of natural vegetation took on added significance. Forests and their soils became important because of their large area, considerable total metabolism, and high carbon content. Worldwide, naturally forested regions were thought to be about 44 percent of the globe's land surface, according to earlier Food and Agriculture Organization (FAO) data used later by the World Commission on Forests.[8]

The issues surrounding the metabolism of forests were more complicated than they appeared initially, however. Sorting out how to measure the metabolism of a landscape, and dividing the metabolism between the assimilative properties of photosynthesis and dissimilative properties of respiration, proved difficult to impossible. Photosynthesis is localized in green structures, leaves, and stems. Respiration, on the other hand, occurs in all living tissues, leaves, stems, roots, and the decay of all organic matter in soils.

By the early 1960s, my colleagues and I were measuring carbon dioxide concentrations continuously on a 450-foot tower on the laboratory site and had discovered quite by chance that during nocturnal temperature inversions, which are common on Long Island, the cooler, heavier air near the ground is stable for many hours. Under those conditions, carbon dioxide emitted from the respiration of the entire forest, plants and soils, accumulates. The rate of accumulation defines the rate of metabolism of soil and plants together, the total nocturnal metabolism of the forest. It was an astonishingly simple measurement and revealing—the first such measurement of the metabolism of an entire forest. Since then, far more sophisticated techniques have been developed for continuously measuring carbon fluxes through the entire depth of the forest canopy.[9] But the early measurements at Brookhaven were enough to show how important forests are in determining the composition of the atmosphere locally and, presumably, globally. After all, the metabolism of the forests was reducing the mean carbon content of a hemisphere by several ppm in a few months … and then restoring it in a similarly short time. Such a change in a hemisphere involves the absorption and release of billions of tons of carbon into an atmosphere that then contained about 750 billion tons.

The evidence of the significance of the global vegetation, especially forests, in determining the composition of the atmosphere was slow to be accepted, even by the scientific community. The reluctance to accept the biotic influences was strange, particularly in view of the conspicuous annual hemispheric oscillation in the composition of the atmosphere. But the importance of the oceans in determining the composition of the atmosphere seemed overwhelming to many scientists. Climatologists and oceanographers of the 1960s and 1970s had cleverly used radioactive isotopes from the bombs to define the absorption of carbon dioxide from the atmosphere by the oceans. It was and remains a significant flux. The focus on the oceans drew attention away from the terrestrial biota, its size and influence.[10] Ecologists gradually became more experienced in sorting out details of the metabolism of the natural communities on land, especially forests, and in appraising the role of those communities in determining atmospheric composition at any moment. Now it is recognized that terrestrial vegetation and the oceans absorb about equal amounts of the carbon emitted by human activities, although the flows are complex and awkward to measure.

A large step in this direction was progress at Brookhaven in defining the net carbon flow of a forest as "net ecosystem production" (NEP). Under some conditions, especially during summer, photosynthesis dominates, and carbon flows from the atmosphere into storage in plants and soil. NEP is positive. Under other circumstances, when light is low and temperatures may also be low as in the northern winter, photosynthesis diminishes and respiration dominates. NEP drops. It may become negative if organic material decays and carbon is lost to the atmosphere. The direct influence of temperature on photosynthesis is small, but its influence on rates of respiration is high. Suddenly it became clear that increasing temperatures might be expected to favor respiration over photosynthesis, especially the respiration of decay organisms in soils, and release more carbon. Such an effect over a large area would put more carbon into the atmosphere and contribute to a further rise in temperature. A significant warming might increase the total respiration in a large region such as the Arctic or the circumpolar boreal forest. Such an increase in places where large pools of carbon have been stored in plants or soils could be large enough to affect the burden of heat-trapping gases in the atmosphere significantly—a potentially self-amplifying change.[11]

As we continued to explore these topics it became clear that much of the respiration of decay in the higher latitudes occurs in bogs, swamps, and soils where oxygen is limited. (In some of these bog soils decay may be further limited by high acidity, even in the tropics.) The immediate gaseous product of anaerobic decay is methane. Over thousands of years methane has slowly accumulated in frozen soils and sediments. Some has accumulated in crystalline form as clathrates in coastal waters. Such accumulations are vulnerable to release as a gas as soils thaw and coastal waters warm. The total amount available from these two sources, soils and coastal waters of high latitudes, is unknown but potentially enormous in proportion to what is present in the current atmosphere. Methane's greater capacity for absorbing infrared energy than carbon dioxide makes it a potentially serious additional contributor to climatic disruption. Contemporary net releases from these sources remain small in proportion to the approximately eight or nine billion tons of carbon released annually by burning fossil fuels. Nevertheless, a rapid warming with increased thawing and drying of the Arctic and boreal forest zones has the potential to add enormously to the contemporary releases from burning fossil fuels and from deforestation.[12]

* * *

There has never been much doubt that adding heat-trapping gases to the atmosphere will warm the earth. Nor has there been much question as to the pattern of distribution of heat as the earth warms. There will be little change in the tropics and substantial changes in polar regions, where the accelerated warming may trigger a cascade of other changes. The biotic reservoirs of stored carbon are especially large and vulnerable in the higher latitudes of the Northern Hemisphere, the very regions where the warming is greatest.

Experience over decades has led to the following widely accepted broad conclusions about the effects of the accumulation of heat-trapping gases in the global atmosphere:

- The warming will be most intense in the continental centers, warming them and, as they warm, drying them out.
- The global continental arid zones will expand as the climates warm—a series of changes especially important in the Northern Hemisphere's large continents.
- The hydraulic cycle will be intensified as energy is absorbed through the evaporation of water in the moist tropics where the energy density is high. The water vapor, carrying the energy of vaporization, enters the normal atmospheric circulation and is carried toward the poles. In the higher latitudes of both hemispheres the vapor cools and condenses, yielding its heat. The warming of the earth will be greater in those latitudes, a particularly significant issue in the Northern Hemisphere because of the large land area in forest and tundra and the large stores of carbon held there.
- Precipitation will increase, especially at the cooler continental margins.
- The increase in energy in the atmosphere feeds larger and, possibly, more storms of greater intensity.[13]
- Glacial ice globally will melt with rising temperatures. The potential if all glacial ice disappears is for a sea level rise of 225 feet, inundating all the coastal cities and towns worldwide. (Ten to fifteen thousand years ago, the sea level was about 300 feet lower than it is now.)[14]

These are not unique or novel insights. They are merely a summary and recitation of perspectives long held by climatologists and other analysts. The insights can be ignored, but at great cost as the capacity of the globe for supporting all life erodes. The evidence should be the basis of never-ending embarrassment of current governmental leaders for it is a derogation of duty—an egregious ethical lapse—that allows such systematic corruption of human welfare globally.[15]

Conspicuous climatic disruptions are already under way well beyond the increasingly acute aridity of continents. As anticipated, the climatic disruption has warmed the Northern Hemisphere differentially. The Arctic has warmed far more than any other region, and summer ice in the Arctic Ocean has been reduced to the lowest areas ever measured.[16] The reduction in the area of ice shifts the balance between reflective white ice and the dark, energy-absorptive water surface. Just how that shift will affect the climatic circulations of the Northern Hemisphere has been beyond experience and capacities for prediction. Yet the energy added in those high latitudes appears to be changing the circulation patterns to producing new extremes of climate in lower latitudes. The extremes appear as anomalies such as storms of unusual size or power. On the US eastern seaboard, Hurricane Sandy of October 2012 was a recent insult—a storm of great size and fury that brought an unprecedented tidal surge that washed away houses and beaches along the New Jersey and New York shores, flooded a segment of the New York subway, and cut power to sections of New York for weeks in some localities. As with the earlier devastation by Hurricane Katrina of New Orleans and many miles of the Gulf coast of Mississippi and Alabama, the damage was so severe that restoration cannot occur. The houses and, in some cases, the land itself, have disappeared and can never be restored. The city of New Orleans, partially restored, remains below sea level, and although protected by dikes, is not in the eyes of many onetime residents a sound risk for habitation in a period of increasingly violent storms.[17]

As troublesome as such storms have been, new records appear year by year. Typhoon Haiyan, which cut a broad swath across the southern Philippines in early November 2013, appears to have killed as many as ten thousand island dwellers and left many miles of devastation. It seems to have been a storm of unusual ferocity—the worst ever to hit the Philippines, according to some reports. The rash of tornadoes in the central eastern states of the United States in recent years lends additional credence to the view that such large storms are increasing in frequency as the climate creeps well beyond normal ranges and far from control.[18]

The damage from such storms is in each case only partially repairable. A fraction, sometimes a large fraction, is simply accepted as the new normal, a new public cost, an increment of erosion of common property resources. As serious as such cumulative losses are, they are trifling next to the potential for triggering irreversibly destructive trends. The warming of the southern oceanic waters, for instance, is currently speeding the decay of the West Antarctica ice cap, which is sliding more rapidly than ever before into the sea. The ice cap is large, and if totally melted, will raise the sea level globally by an estimated ten feet or possibly more. But other glaciers are melting, too, including the large Greenland ice cap—another twenty feet if totally melted.[19] The rise in sea level over the next decades could easily be ten feet or more. The storm surge of Sandy that flooded a New York subway was about ten feet. Much of southern Florida is vulnerable to a ten-foot rise in sea level or even an equivalent storm surge. So is every delta globally.

Even worse is the potential for the warming to trigger large emissions of additional carbon as carbon dioxide or methane from melting Arctic soils, thereby taking out of human hands any possibility for avoiding a climatic catastrophe. The public is becoming aware that the world is on the edge of what could become a terrifying cascade into climatic chaos.[20]

* * *

The question of what to do—the great issue of these early decades of the new millennium—is explored in the following chapters. There seems little question as to the needs of a seven-billion-person world rapidly moving toward a world of ten billion or more. The obvious need is a stable and predictably supportive biosphere, the antithesis of the climatic disruption outlined above. Such a return to stability is still possible with a concerted effort, but it requires a return to an atmospheric burden of carbon dioxide approximating that of 1900, 300 ppm or slightly less. Such an objective does not embrace the popular compromise that a two degree rise in the average temperature of the earth would be acceptable or in any context "safe." The "350" objective set forth by Bill McKibben and his colleagues is a fine interim goal, but does not offer long-term security unless it marks the halfway point in a rapidly moving trend toward 300 ppm or less. The problem is that 350, if achieved, while infinitely better than the more than

400 ppm we have now, would still constitute a large warming of the earth with all its associated impoverishment of biotic systems at a time of soaring human needs.

The conspicuous difficulties in obtaining joint action by 193 nations in making drastic reductions in the use of fossil fuels has encouraged speculation as to what extraordinary technical steps might be taken in an emergency to reduce global heating. Such "geoengineering" might involve efforts to reduce heat-trapping gases rapidly or attempting to change the reflectivity of the earth as a whole through interventions in the physical properties of the atmosphere. The removal of heat-trapping gases, especially carbon dioxide, is obviously necessary. Most geoengineering proposals, however, involve tinkering with earthly factors such as the reflectivity of the atmosphere—changes that could have profoundly disruptive, unforeseen, and unacceptable biophysical side effects. Biologists and other earth scientists almost universally scorn the idea of reaching outside the normal factors that stabilize the biosphere to invoke short-term solutions through massive changes in the atmosphere or oceans, seeing that as unjustifiably dangerous.[21]

That spectacular view of the Amazon on the front page of the *New York Times* called to mind my own many visits to the Amazon basin of past years. There, the story of human survival and climatic disruption takes a different, more urgent cast. People have built lives in the Amazon basin for thousands of years. Families of the *varzea*, for instance, the seasonally flooded marginal lands of the river, earn subsistence from the river, its periodically flooded marginal flatlands, and from the uplands nearby. Some migrate with the flood to the uplands, and manage crops and livestock in both places. Some live with livestock on rafts during the flood and return to the land for pasture as the river recedes. These families are integral to the varzea and have been for generations. Their self-sustaining independence is threatened by the encroachments of modernity, industrial fishing, and the giant farms that suddenly bite large tracts from the forest and force people off the land.

And then there was the energetic young farmer from the Amazon whom I met much later when he came as a visitor to our institution. One could not but be caught up in his enthusiasm and plans for managing his family's giant soy-producing enterprise, protecting streams and carefully avoiding erosion of higher ground, fertilizing soils and managing giant combines for harvests—exciting, sensible methods and bold objectives. He lives a common dream, travels in the Amazon by small plane, his own, hopping from one part of the large family business to another. He hires local talent to handle crops which go to international markets where he is also an expert and participant. He told us that life is dangerous. Some object to such use of land until recently occupied and used by others, indigenous families possibly even the varzea dwellers. And there are others, competitors, who might also aspire to building large agricultural businesses on "new" land in the Amazon. Laws are weakly supported on the frontier. He carries a gun.

But the issues remain, framed starkly by that 2009 photograph showing the division of forest from incipient desert. Is there a compromise?

In the seven-billion-person world, there is no middle ground that preserves both the Amazon basin's core role in global biophysics and opportunities for building industrial agricultural empires that displace forests and people. The time now is to celebrate the lives of those now living who find life attractive on long-occupied lands using age-old techniques. The modern world can improve those opportunities and lives, as we shall see, but within the core of essential global environmental integrity. There is no room or time for the industrial agricultural transition of that startling photograph. Global biophysics and the largest human interests now define a different path.

With insight, imagination, respect for facts of the world, and discipline, there are yet ways out of the trap we have set for ourselves—ways that are much closer to celebrating the carefully balanced lives of today's varzea dwellers than the incipient deserts of a misplaced industrial frontier so conspicuously displayed in the *New York Times*.

II ENVIRONMENT IS POLITICAL: CLIMATE HEATS UP

THE GLOBAL COMMONS:
A CORPORATE FEEDLOT

Profits can be increased by keeping wages low and real social, environmental and economic costs externalized, borne by society at large and not by the firm.

—James Gustave Speth, *America the Possible*

E. Cuyler Hammond, a talented scientist and statistician whose experience I was fortunate to be able to draw on in the early 1960s when I was designing the forest experiment described in chapter 2, had ten years previously helped to develop a detailed study for the American Cancer Society of the health effects of smoking. It was a magnificent early demonstration of the potential of how the new punched card computer techniques that IBM was developing at the time could be used in accumulating, managing, and interpreting large sets of data. The carefully controlled epidemiological study compared factual observations of the health of thousands of smokers with the experience of nonsmokers. The data were devastatingly clear: smokers lived shorter lives and more frequently suffered from cancer and heart disease as well as other serious afflictions.[1]

One might have at the time, 1958, expected that a carefully designed and executed comprehensive study by talented scientists would be widely acclaimed and accepted as definitive. Far from it. The tobacco business was entrenched, rich with money, and supported by millions of smoke-addicted citizens. Its tentacles reached far into agriculture with carefully developed and jealously guarded governmentally assigned allotments for the production of tobacco, a cash crop of great value. One study, no matter how powerful, was a mere whiff of passing fresh air, clearing the thick haze for a second before the smoke closed down tightly around the tobacco business, which only expanded over the next years. Recognizing the dangers of such studies, the cigarette companies moved powerfully to defend themselves with liberally financed new "scientific" studies of their own, designed, if not to deny the hazards of smoking, to obfuscate and cast doubt on the health effects, and deflect any legislation or attempt at regulation. Scientists, some of them, were pleased to take money to contribute to this transparently corrupt charade.

The recent decades of the tobacco story have been delineated and lucidly documented in *Merchants of Doubt* by historians Naomi Oreskes and Eric Conway, and independently in *Golden Holocaust* by Robert Proctor.[2] These books not only put the lie to industrial and commercial claims of propriety in dealing with the public interest and welfare but also show how the tobacco industry's response to the scientific challenge posed by studies such as that of Hammond and Horn has been adopted as a model for corporations in protecting profits against private lawsuits and governmental intervention.[3] The key to corporate success in this realm was to question the data's validity, express serious doubts about earlier work (regardless of its quality), and assert that unless further studies were done, the data were too uncertain to warrant any change in company behavior, regulation, or use of the product. In the case of smoking, the industry managed to delay restrictive action for fifty years after the American Cancer Society's study. The industry had, in effect, established the right to kill with impunity, profit along the way, and had devised a strategy for extending, even expanding, that right and their profits.

I was unprepared for these distortions as I encountered them with DDT, first in Maine and then on Long Island. I was puzzled at their reality and persistence in all realms, not just commerce, but in academic science and government as well. I learned of them firsthand in the pesticide discussions where commercial products such as DDT brought corporate money for research to the state university system, which responded cooperatively. My perspective grew increasingly critical, honed by the realization that the spraying of the forests of Maine and New Brunswick was perpetuating the pest problem along with the need to spray annually. Research, a special realm of the agricultural university, was focused on commercial chemicals for pest control as opposed to cultural techniques or biological controls that may be more complex but never enter the market. Science, nominally a search for truth, took an industrial-corporate bias that favored commercially attractive methods, and scorned, or at least neglected, the search for biological controls that might be enduring, safer, and inexpensive. Later, as a staff member in a national laboratory operated under the Atomic Energy Commission, I realized that we, too, were making continuous compromises with human safety and well-being in handling radioactivity. The whole new industry centered on the atom depended on such compromises in both construction and operation. And it mattered little whether the objective was weapons of war or a new source of energy for the world.

Meanwhile, the issues of human rights and welfare drew the attention of the quite-unusual scientist, Garrett Hardin. The issues were becoming acutely intensified by soaring human numbers and expanding industry and technology. Despite the excitement of Hardin's original focus on genetics, a critical look at a world closing in around us all led him to put his full efforts for the latter half of a highly productive academic career into analyses of the implications that he had developed lucidly in "The Tragedy of the Commons."[4] Hardin defined the tragedy as the erosion of those elements of environment shared by all: air, water, and land and civil rights. He saw the corruption of public resources by private interests as infringements of human rights and welfare. Civil rights crumble as corporate and other private interests pursue profits by pushing costs into the public realm. Hardin was quick to observe that the protection of common property and equity in all matters requires rules, mutually respected. He had no time for popular assertions in the business world that the free market system would ultimately protect all such resources and rights. Similar sentiments led David Orr, a political science scholar at Oberlin College, to the blunt contention that "all education is environmental"—a further acknowledgment that civilization requires a fully functional environmental system that sets limits on political and economic adventures.[5]

Democratic capitalism has been widely accepted in the West as the most successful governmental system. Nevertheless, its success has entrained devastating environmental excesses. Among contemporaries, two major figures, Gus Speth in *America the Possible* and Al Gore in *The Future*, have recently embraced Hardin's critical perspective that aspirations for corporate growth at any cost devour civil rights and often welfare in general. They condemn corporate expansiveness and greed in reaching for control of science and government. A public understanding of the necessity for preserving the functional integrity of environment universally is essential, they argue, and requires relentless attention and demanding effort to avoid sliding into a narrow economic model subservient to profits at any cost.[6]

As though specifically scorning that perspective, in early March 2013 the *New York Times* announced that it was closing its environmental desk. To be sure there were financial issues. Newspapers are having a hard time in competition with electronic media, and environmental news in this new corporately dominated world was not selling well. Yet it was a troublesome step for it appeared to leave the core of the news to be defined by the market as politics and economics, flourishing agents of the giant corporate interests that have presented the world with the clutch of environmental bombs, fuses lit and short, that are the topics of Speth and Gore as well as a score of other treatises including this one. Snuffing those fuses requires clear insight into causes and consequences, long continued, and supported by intellectual insights. The *New York Times*, financially pressed, decided not to lead in that direction but rather to follow the momentum of industrial growth.[7] That momentum conveys an industrial bias to the news that is difficult to avoid unless recognized and countered deliberately. The press may be free but it, too, follows the money. But the core of the news is environmental.[8]

The dimensions of the challenge in the United States in visualizing an honest and fair appraisal of the news emerge from the magnitude, implications, and irreversibility of the chain of recent environmental disasters. All were and remain products of corporate exuberance that should have been tamed in advance by governmental controls and public insights cultivated by a well-informed, probing discussion. In retrospect the projects fed short-term business interests while destroying only slightly longer-term interests in human safety and welfare. They include the Fukushima catastrophe, the Gulf oil spill, the climatic disruption, the abandonment of Alberta to the mining of oil, sale of Wyoming, Montana, and large pieces of the southern Appalachians for coal, the expanding rash of toxic wastelands over the Marcellus Shale as gas and oil are extracted by "fracking," regional and global pollution by tens of thousands of chemical wastes, and other brazen corporate intrusions. Underlying all is often the dismally disappointing mantra that the free market system has all the solutions needed in a world large enough to dilute virtually any intrusion.

The Fukushima disaster occurred on March 11, 2011. One of the world's politically most stable, well-organized, and by comparison with many nations, wealthy industrial societies fell into chaos as the tsunami killed thousands and destroyed three of six nuclear power reactors, which spread poisonous radioactivity over hundreds of square miles of a densely occupied island nation. No amount of political or economic manipulation could do as much as define the problems with the plants, much less quickly resolve them. The tsunami itself was an act of nature, but its consequences at Fukushima need not have been a serious nuclear disaster.

The news two years later in the industrialized world was that a high-stakes gamble had been taken and lost by those entrusted with the public welfare in Japan. It was, at least nominally, a series of joint decisions between governmental and corporate interests to build the reactors on an exposed Pacific coast. There were industrial profits to be made from selling energy into a public market in an otherwise energy-poor nation. The convenience of the site, size of the market, and needs for power dominated the decision that generously favored TEPCO, a corporation whose business was selling power. Environmental considerations, not at all subtle and widely understood, if publicized, would have revealed the extraordinary risks to the landscape and the public in consolidating regional energy production in six reactors in one place, and especially on an exposed Pacific coast, obviously vulnerable to a tsunami or serious storm. Worse, TEPCO was well aware of experience with other reactors and had not taken precautions to avoid such a disaster.[9] The faults here were multiple: governmental, industrial, and informational. Again, the incentives were inadequate, the information was incomplete, and the public and the public interest were misinterpreted or ignored.

The risks were clear enough in advance. Six reactors close together at sea level obviously compounded risks. Proximity to the sea provided easy access to cooling water for the power plants—but raised the possibility of flooding in a storm. Worse, a serious accident in any of the six would threaten access to, let alone the integrity of, the others. And then the hazards to thousands of residents nearby were to be considered. Governmental oversight obviously failed; yet so did all the various potential watchdogs including the news media. It is difficult to believe that an informed public with access to a thorough environmental review, and with a government sensitive to the public welfare, could have allowed the construction of six reactors in that place.

Half a world away in the Ecuadorian Amazon a different, more common twenty-year struggle continued between the government of Ecuador and the international industrial giant Chevron Corporation. An Ecuadorian court ordered compensation to be paid to thirty thousand forest dwellers in the Amazon basin for extensive contamination and destruction of an area the size of Rhode Island by Texaco between 1964 and 1990 in drilling for oil. Chevron now owns Texaco, and the court found the company liable. Chevron simply denied Ecuadorian "sovereignty" and invoked various trade agreements that it alleged protect such companies from damage claims. Similar industrial invasions of local human rights have followed corporate mining in Patagonia, Chile, and the southern Appalachians of North America. All stand as cautionary tales for nations that look to large corporations as a source of economic development. Further, international trade agreements are specifically designed to offer corporations the ability to undermine restrictive national environmental laws as well as labor laws and interests.[10]

The problems of Japan and Ecuador are but a small sample, part and parcel of the intensification of human occupation of the earth. Local residents of these two nations are victims of industrial civilization with all its technological and social advantages. Japan, a wealthy nation with substantial political and economic resources, will recover over time, although many citizens will suffer. Ecuador will suffer for years, largely ignored. The radioactive landscapes of Japan cannot soon be reoccupied for farming or habitation with safety assured for all including customers of farm products. Environmental support of the political-economic systems has been seriously weakened. It has been destroyed for the approximately hundred thousand residents displaced from the heaviest fallout zone extending twenty-five miles northwest of the plant on the main island of Japan.[11] "Cleanup" is an attractive phrase, but there is no possibility for removing long-lived radioisotopes from land and sea apart from waiting for radioactive decay to diminish the hazards. The pressures to reoccupy the land will soar and the new hazards to health will appear at that moment to be remote in time and vague by comparison with the needs to use the landscape. The vague and often long-delayed costs will be accepted and borne by those who have little choice, who cannot move elsewhere and thrive. The impoverishment is real enough but it is simply incorporated into society as another burden—a corporate cost shared unwittingly by the public.

Then there is the marine contamination, which was large, possibly the largest such release ever.[12] While the radioactivity released into the coastal ocean becomes invisible immediately, it can reappear in surprising places as it, too, persists, and is passed through currents, chemical transitions, and food webs over years to emerge in the tissues of fish, birds, potentially people years later. It is a heavy cost—a largely invisible one—which will linger as a hazard to life for decades and remain recognizable as isotopes in the geologic record substantially forever.

* * *

Decisions such as the siting of reactors in Japan and the commitments of forest and land to oil extraction in Ecuador may appeal as wise at the moment, ignoring vague and distant hazards, but they were both made by corporations that stood to profit. Such decisions belong in the realm of community governing bodies acting in the interest of the public with full disclosure of the hazards. If the consequences of a catastrophic failure are unacceptable, even a low risk of failure becomes unacceptable. If one cannot afford to lose a bet, don't bet. There are alternatives to building reactors in places vulnerable to tsunamis and to multiple nuclear disasters, ... and alternatives to oil extraction in the richest forested areas in the world. The expanded human presence and intensified use of the landscape requires greater attention to guarding common interests, not less.

In direct opposition to governmental regulation sit those who persist in belief that the free market system, lightly regulated, will define human needs, ration resources equitably, and protect both resources and human rights. The concept is neither new nor original. It is a popular political stance for business including industries that find their activities limited by rules protecting civil rights or public access to common resources, such as clean air and water. That somewhat anomalous credo gained enormously in the United States on Ronald Reagan's election, when he announced repeatedly that "government is the problem" and that as president he planned to "take the government off the backs of the people." That mantra has been adopted and amplified by a vigorous political right in the United States over the thirty-five years since the 1980 election into a major effort to deny the reality of the climatic disruption and the propriety of governmental regulations protecting environmental quality. The general attitude was epitomized by Steve Forbes, president, CEO, and editor in chief of *Forbes Magazine*, in 2007, in objecting to my concerns about the disruption of climates globally: "Weather patterns have been constantly changing for aeons, long before humans inhabited the earth. It is scientific hubris to conclude we truly know what affects those patterns and how."[13] Such hostility to science and antipathy to governmental protection of public resources and welfare have developed powerful political support. The transition away from the intense and bipartisan interest in environmental stewardship of the 1970s, which produced the highly professional congressional Office of Technology Assessment (OTA) in 1972, began in 1980 with the Reagan era, and has brought overt efforts over more than three decades to cripple governmental capacities for doing its job in regulating well-known poisons and protecting all aspects of common property. The OTA was a valuable source of objective insights into the full range of environmental hazards, although it had to walk softly to avoid shrill opposition from industries and their political partisans. The opposition from commerce and industry, all feeding on the commons of time and space, air and water and land, led to its closure in 1995, cutting off Congress after twenty-three years from its own source of objective scientific appraisals. In closing the OTA, Congress was obviously yielding its duties and loyalties to its wealthiest corporate clients. The move stands as one of the most memorable contributions in recent times to the celebration of ignorance in government, and the potential for corruption of public purpose and welfare.[14]

The corruption is contagious. Chevron in the Amazon simply turns its back on Ecuadorian governmental authority. Industries commonly work to avoid such responsibilities item by item, challenged only after the fact, and then denying even powerful proof of the invasion of human rights and health by contamination of resources shared by all. The general assumption is that the commons are open, fair game for all unless protected by public irruption and action by a large, powerful government. BP, unable to escape, yielded to the US government in the Gulf and was forced to pay huge reparations, but won in the end: the drilling continues against strong objections from scientists, residents, conservation agencies, and common sense. Commercial purveyors of pollutants, lead, mercury, sulfur, nitrogen, insecticides, herbicides, fungicides, plastics, hydrocarbons, antibiotics and medical products, "cures" and wastes, all enjoy a public subsidy in the form of irresponsible releases into the general environment without cost or, in many cases, even accounting. The scientific, technical, commercial, and medical regulatory interests battle continuously over the effectiveness, safety, and rights to sell and profit from monopolies on scores of drugs, chemicals, and other substances nominally regulated by government. Each is considered unique, and if environmentally hazardous or threatening, must be proven so to be contained. Once proven toxic, efforts continue to defend a product's use in low quantities or for special circumstances. Experience with DDT and radioactivity, definitive as it was and remains decades later, is taken not as the basis of precautionary rules but as singular events, interesting but of no larger consequence. The whole series of corporate arguments is a routine, repeated for every new commercial venture that gains from the sale of hazardous wastes that ultimately drift into a public trough. The procedure is ludicrous, but it is common corporate credo, a corrupt perversion for profit. By that set of popular industrial procedures, science is never predictive, never adequate, never prescriptive or appropriately invoked on the basis of experience. The shameful legacy of Reagan has been exploited to cultivate a significant political bloc that works to ease governmental regulatory activities even to the point of eliminating the EPA.[15] It is a strange stance for any citizen to endorse the corruption of essential commonly held resources—a part of every American's heritage to be enjoyed, and then passed on as a legacy to children and all other successors. And for public representatives in Congress entrusted with the welfare of all to assume such a stance is corruption beyond contempt.

The perversity of the Reagan doctrine, carried to extremes, is obvious enough. It becomes doubly so when adopted as the policy of advocates of economic "growth." Growth, as these partisans see it, usually implies larger markets, access to more people, further industrial development, and more commerce in general as well as increased personal wealth and, presumably, welfare. Corporate interests regularly advocate such development, sometimes even seeking to increase their contributions to environmental corruption in the process. A California court decision in 2013, for instance, assigned a heavy fine ($1.1 billion) to three companies for deliberately selling lead-based paint, despite having full knowledge that it affects brain development in children and brain function in adults. Similarly, General Motors Corporation deliberately concealed its knowledge that automobile ignition switches were failing and causing serious accidents.[16] As economic expansion is pursued and corporate growth proceeds, new points of conflict require rules and standards of behavior that lubricate all aspects of activities. Such rules are the core purpose of government. Yet those nominally defending the "rights" of individuals, business, commerce, and industry vigorously scorn these rules. Unfortunately, the free market puts the commons, all public resources, at the mercy of exploiters. Again, the tragedy of the commons applies until rules are established, agreed to, and respected by all.

Forgotten in these discussions is the history of unregulated market systems that have often dominated commerce and much of human affairs for centuries. Civil rights for all citizens did not emerge full blown and well defined as well as respected from unfettered commerce. They were the result of hard-fought struggles over centuries, and are defended even now but imperfectly. In an increasingly populous world with diminishing resources, the struggle will be ever more intense. If we continue to allow the inversions of responsibility in industry and commerce that systematically compromise the environmental commons for profit, it will be a wretched world.

* * *

It is not surprising that the corporate interests of an industrial nation would find it attractive to build six reactors adjacent to one another despite vulnerability to a disastrous flood. Nor is it surprising that a giant oil company could substantially command the whole of the Gulf of Mexico by losing a series of risky bets in drilling for oil in deep water. These are incremental steps in the environmental corruption of nations. The costs accumulate on a global scale, and with some glittering exceptions, the national wealth and human welfare suffer. Where does the process lead? Does it have an end?

We do not have to look far to find examples of what pushing such environmental corruption to extremes generates. Japan itself, its disaster still fresh and continuing, but its government intact with many resources at its command, is recovering. The nation, hesitant about an earlier decision to lean heavily on nuclear power, began some serious renewable energy projects, managing within months to harness wind power equivalent to the output of two modern reactors.[17] But national environmental disasters fester in Haiti, Sudan, Somalia, Syria, and Mali, and lesser disasters are brewing in Mexico, India, China, and a dozen other states—all examples of misguided governments, a market system run wild, and an environment exploited without rules to the point of gross impoverishment.

Superimposed on already-crumbling societies and pushing others toward the brink of collapse are the mounting burdens of a global climatic disaster brought on by rising fossil fuel use. A world environment that historically was large enough to be little affected by centuries of human activity, at least at the global level, if not always locally, is coming to have a definitive role in determining global human welfare and collective wealth.[18]

The devolution of the environment is conspicuous in the increasing chaos of nations as well as migrations of millions in and out of Africa. The unrest in Africa is not surprising. The region is becoming increasingly arid as the climatic disruption proceeds.[19] The fifty-one sub-Saharan nations currently have about 900 million inhabitants, and that number is expected to swell to 2.2 billion by 2050, constituting the largest population growth in the world.[20] The unrest in Somalia and elsewhere all have a large environmental component with water shortages and crowding as large factors. Pressures to migrate to stable nations in Europe are high and growing. The common routes in the west are from points in northern Africa by crowded, unsafe boats to Sicily or mainland Italy, and to Spain from Morocco or the Canary Islands. It is a lucrative and dangerous business because people are willing to give their life savings to enter Europe. The numbers attempting the crossing are in the tens of thousands. For southern Europe, a BBC report estimated illegal crossings for 2009 and 2010 at about 105,000 each year. In the year of the Arab Spring, 2011, it jumped to 140,000, then dropped in 2012 to about 70,000, and rose again in 2013 to more than 105,000.[21]

The increased aridity of Africa and other continental centers is a product of the totality of industrial development elsewhere that receives a magnificent subsidy in being allowed to poison the world with its fossil fuel wastes. The corporate world as a whole now commands extraordinary wealth and political power, remains heavily involved in developing and using fossil fuels, and is reluctant to consider any alternative. Giant corporations and business interests have become so wealthy and politically powerful that they displace popular democracy, rewrite laws to favor their own interests, and sow fear among the public to prevent objections. Thus it is that political and economic controls slip into self-serving hands. The transition is conspicuous in the United States, but it reaches into the international realm as the collective effects of giant corporations become large enough to wreck nations remotely with impunity.

The most successful governmental system in the world, the epitome of democratic capitalism, is vulnerable. The US Supreme Court's finding in the *Citizens United* case removed limits on corporate funds in support of candidates for public office, thereby greatly strengthening corporate political power in the nation. The decision brought an unprecedented flow of money into politics from widely diversified corporate interests, intensifying congressional scorn for facts. The fossil-fuel-producing industries and their allies of course favor politicians and laws protecting the status quo: a flow of profits despite the progressive poisoning of the world by their waste products. The importance and magnitude of that interest became apparent early in the second election campaign of Barack Obama. The *New York Times* reported that "two months before Election Day ... estimated spending on television ads promoting coal and more oil and gas drilling or criticizing clean energy ... exceeded $153 million ... [or] nearly four times the $41 million total spent by the entire Obama campaign and Democratic groups, clean energy advocates, and others defending the president's energy record or raising concerns about global warming and air pollution."[22]

The United States is not a lonely target. The potential wealth from the tar sands of Alberta has captured Canada's Harper administration, which is unabashedly bending all efforts to mine and sell tar sands oil to the world.[23] The Harper administration has also circumvented scientific critics by closing environmental research laboratories and by firing scientists who have worked for years for public agencies developing information on the careful management of Canadian land and water resources for sustainable use in the long term.[24]

The money involved in the mining and sale of coal, oil and gas is proportionately of even greater consequence in less industrially developed nations. In 2007, for instance, Ecuador's president, Rafael Correa, in the words of the *New York Times*, "offered the world what he considered an enticing deal: donate $3.6 billion to a trust fund intended to protect nearly 4,000 square miles of the Ecuadorean Amazon and his country would refrain from oil drilling in the rain forest."[25] The plan was to be administered by the UN Development Programme and would have preserved Yasuni National Park from invasion for oil development.

Despite an enthusiastic response from conservation interests, not surprisingly governments failed to embrace the plan with major contributions, and the burden fell primarily to Ecuador and international conservation interests, which managed to come forth with but $13 million. Correa ultimately announced that without adequate funding, the plan could no longer be allowed to stand in the way of the "wealth" promised the nation from oil development in the basin. Ignored was the fact that such wealth rarely flows to the nations involved, but rather to the corporate interests and, often enough, local politicians. The nation is left with empty coffers, depleted oil wells, poisoned water supplies, and the toxic residue of the operations, all gratuitous contributions to the erosion of common property resources. The Park is public property and its protection is the responsibility of government in the public interest.

Many other nations face the same tempting bait: sell a rich, biotically renewable land surface of limited immediate economic potential for the promise of immediate wealth for some from mining of one sort or another. The short-grass plains of Wyoming, long preserved for grazing cattle, and for centuries before that grazed by the North American bison, are currently being destroyed in the interest of selling the deep coal seams underlying them. The coal is sent to China in support of industrial development, even as the wastes accumulate in the global atmosphere. Meanwhile, profits are funneled to corporate owners, and the costs are diffused into the public realm as wasteland, the acceleration of a global climatic disaster, and the consequences of the distribution of directly toxic products, including soot and mercury from the combustion of coal.

These developments are examples of the free market system's normal operation. Political influence follows the accumulation of wealth to further what Sheldon Wolin calls "inverted totalitarianism."[26] The political and economic worlds are inverted with corporate interests displacing governmental function to favor corporate purposes. Affected are civil rights, the control of money and jobs along with land use, and environmental affairs as well as human rights. While the industrial development promoted may offer amenities in employment and wealth for some during the harvest, the relentless destruction of renewable resources accumulates as a public burden of biotic and environmental impoverishment. The process has many facets. It includes not only biotic impoverishment but, far too often, pervasive and persistent economic and political impoverishment as well. Lester Brown, agricultural economist and founder of the Worldwatch Institute, has produced a wealth of examples of the slide of nations into impoverishment. He tabulated "failed" and "failing" states in the first decade of the new millennium, borrowing an earlier tabulation that appeared in the July–August 2005 issue of *Foreign Policy*. The criteria of "failure" involved the loss of governmental control over "part or all of their territory [to the point where governments] can no longer assure the personal security of their people." Further decline frequently entails loss of ability to collect taxes or provide for public services, including those essential to a functioning economic system.[27]

Twelve criteria were used to rank sixty nations in varying degrees of failure. Examples of such nations were, not surprisingly, abundant. The top five were in Africa with Somalia heading the list as the most thoroughly depressed. The only New World nation was Haiti, eleventh on this scale. Haiti stands, grossly impoverished, on the threshold of the United States, alone and substantially ignored in the Western world. The current malaise there is not a result of climatic disruption, nor is it from the routine operations of vast corporations, although they have had a hand. The impoverishment of land, water, and the citizenry in Haiti stands as the cumulative product of incremental biotic impoverishment—a classic example of what in other ways may be in store for many other nations as population growth continues and climatic disruption gathers momentum, unbridled industrial exploitation eats the commons, and essential resources crumble below local capacities for repair. The United States, despite its quite-different history, and its size, wealth, and promise, would not be immune in a deteriorating environment to a slide in this direction. As an illustration of the depths of losses possible as increments of impoverishment accumulate, the Western Hemisphere has Haiti and its long history of exploitive commerce.

* * *

Modern Haiti is the product of a five-hundred-year history of political and economic laissez-faire. The western end of the large island that Christopher Columbus colonized for Spain in 1492 and called Isla Española ultimately became known as Haiti—probably derived from an Arawak name sometimes written as *Ayiti*. The island was a tropical dreamland, easy to live on, and European colonies there became wealthy on sugar and coffee. That wealth was fed by slaves, imported in large numbers from Africa over more than two centuries and under the most miserable of conditions. In 1789, the total population was about 520,000 including 450,000 slaves. The French Revolution inspired a local revolution that abolished slavery and in 1791 established a new nation, three hundred years after the first Europeans came ashore. History did not favor the nation, however. A century and a half of political and economic chaos ensued with leadership constantly challenged as it bounced virtually year by year from thugs to thugs and war to war. France burdened it unscrupulously with a huge debt for its independence and the abolition of slavery. Early in the twentieth century the United States declared it a US protectorate and occupied it from

1915 to 1934. Once a garden paradise, Haiti now imports a major fraction of its food, much of it through US foreign aid. Its government has been left in shreds.

Dire predictions for Haiti's future of a decade and more past have become today's reality. The population had soared to nearly seven million by 1996. Fifteen years later, it was closer to ten million. It had long ago exceeded the capacity of a mountainous land area of 2.7 million hectares (about 10,000 square miles) to support itself. The growth in population continues with a doubling time variously estimated as between thirty and forty-five years, despite decades of realistic concern among nearly all.

In the late 1990s much of the once-arable land of Haiti had been eroded to rock, otherwise impoverished, swallowed by cities and their expanding slums, or was no longer sufficiently supplied with moisture and available for agriculture. Small-scale agricultural plots, subsistence for some, had even then been pushed onto mountainous slopes too steep for successful long-term cultivation. Forest area had dropped to less than 3 percent of the land and has continued to decline before an insatiable demand for fuel, including charcoal for city markets. With the destruction of forests, water from torrential tropical storms, once retained on forested land, ran off quickly to the sea, resulting not only in a perpetual shortage of freshwater even for domestic use but also in eroded slopes and silted reservoirs, roads, municipalities and coastal waters.

Haiti had passed well beyond the point where the basic assumptions of elementary human livelihood applied. It is a classic example of the erosion of the basic human right to livelihood—the body of rights that we build governments to protect. Corporations, incompetence in government, and the practices of many individuals now undermine those rights.

Ill defined as absolute requirements by either science or government, the rights to biophysical necessities are open to encroachment by others scrambling for survival or profit. The health and welfare of citizens is challenged, if not directly affected, even in the democracies. The process continues, accelerated by corporate excesses, to the late and sad stages of erosion of human status, described by Lester Brown as "unsustainability":

> "The ecological symptoms ... include shrinking forests, thinning soils, falling aquifers, collapsing fisheries, expanding deserts, and rising global temperatures. The economic symptoms include economic decline, falling incomes, rising unemployment, price instability and loss of investor confidence. The political and

social symptoms include hunger and malnutrition, and, in extreme cases, mass starvation; environmental and economic refugees; social conflicts along ethnic, tribal, and religious lines; and riots and insurgencies. As stresses build ... the nation-state disintegrates, replaced by a feudal social structure governed by local warlords as in Somalia, now a nation-state in name only."[28]

And that was the state of things before the devastating earthquake of 2010. Already overwhelmed by poverty along with political, economic, and environmental chaos, any hope of recovery was all but crushed. Upward of one-quarter million died immediately, and as much as one-third of the population was left without shelter. The disaster revealed to the outside world the extent of naked Haitian poverty and vulnerability. Paul Farmer tells the story in *Haiti after the Earthquake*, focusing on the human pathos as seen by one steeped in the details of medicine and public health, desperate needs under such circumstances.[29]

Farmer returns repeatedly to the need for reliably clean water to avoid the exploding catastrophes of cholera and typhoid fever. Despite the squalor of Haiti, cholera appears to have been absent prior to the arrival of UN officials after the earthquake. Clean water is a product, and possibly the most important one, of a functional landscape. It was once abundantly available in Haiti but it disappeared with the biotic impoverishment that accompanied the population expansion and destruction of Haiti's forested drainage basins. Governmental quality declined as poverty became more common and corruption accumulated in and outside government. Farmer's story simply confirms that Haiti is in an environmental, political, and economic abyss, unable to move itself out without massive outside aid and a complex and bold plan.

There have been many efforts over two hundred years since Haitian independence to form a lasting political and economic system for that one-time rich, self-sufficient island nation. All have failed. In every instance there is some special circumstance, some specific reason peculiar to Haiti or to the moment, or to the French colonials or slavery, or the abolition of slavery and loss of "wealth." Meanwhile, the forests disappeared, the rivers flooded, the slopes eroded, the fisheries disappeared, the farmland became slums, agriculture was driven to slopes too steep to cultivate, and water supplies were lost. Those who argue that the free market system is adequate to protect essential resources including human rights, free of governmental regulations, should look closely to determine whether and how such a system might have restored Haiti at some point in those two centuries. Rich businesses based at one time on sugar and coffee failed and were not replaced. No internal business emerged, and no external corporation or combination of corporations appeared to exploit a once-rich landscape and an abundant supply of labor and resources. The landscape was systematically and progressively impoverished, ... and the nation followed to become the most impoverished nation in the Western Hemisphere.

Such slides into impoverishment were the concern, too, of George Perkins Marsh a century and a half ago. In the early pages of *The Earth as Modified by Human Action*, he describes the depredations of the Roman Empire of the early Christian era and before. Marsh wrote that its "fairest and fruitfulest provinces ... endowed with the greatest superiority of soil, climate, and position ... best fitting it for the habitation and enjoyment of a dense population ... are now [so] completely exhausted of their fertility ... [as] to be no longer capable of affording sustenance to civilized man." And he elaborated on the impoverishment of the lands of "Persia and the remoter East that once fed their millions with milk and honey," but had become "too poor in superfluous products and too little advanced in culture ... to contribute anything to the general moral or material interests of the great commonwealth of man."[30]

* * *

The vigor and even survival of nations have ever been coupled closely to intrinsic resources, especially environmental integrity. But the systematic *global* impoverishment of the biotic systems—including forests, fisheries, and agriculture—that support civilization is new. At issue now is a contest among models of the world, worldviews, even as the clarity of accelerating global disruption of climate shakes the world. Any solution to the cascading disruption will come not from the hidden hand of the free market but from deliberate and systematic regulation assuring protection of the commons. Again, astute economists remind us that the free market's primary purpose is profits, not the public good, and requires rules to protect both the market and the public welfare.[31]

Economist Herman Daly in various powerful papers and books has elaborated on the implications, now widely acknowledged, that the economy is a subset of the ecosystem, not the reverse. He showed how worldviews must change as the world moves from empty to full.[32] "[The] human economy has passed from an era in which manmade capital represented the limiting factor in economic development (an 'empty' world) to an era in which increasingly scarce natural capital has taken its place as the limiting factor." In a "full" world, for example, lumber production is limited increasingly by the availability of logs, not sawmills. The fishing industry is limited not by fishing vessels or markets for fish but instead by the depletion of the resource. More technology in this circumstance is not the solution. Better husbandry is, Daly observes.[33]

The transitions from "empty" to "full" quickly slip over the edge of stability to progressive impoverishment. Poverty moves such nations away from major channels of commerce and out of the global public eye, but the biotic impoverishment of land and water is far more common, and far more important than most recognize. Soon, more and more of the world is eating the seed corn.

Now, added to the intrinsic limitations of a full world is the ubiquitous migration of climate out from under all life, including the most advanced industrial agriculture. The Central Valley of California, the giant market garden for North America, was crippled by drought in 2014. The soybean ranches of the Amazon, a symbol of the rich industrial world, are vulnerable as the removal of the forest to found ranches contributes to the increasing heat and aridity of the basin.

In a full world, the purpose of "development" must change. It must change from mere stimulation of economic activity to the reestablishment of a stable, viable landscape in support of a viable economy in a steady state. In that world, economics and politics become means to an end, not ends in themselves. It is the landscape that provides the basic resources as well as the place on which to stand to build political and economic systems within the biophysical limits of the place.

The concept is not new. Settlements from campsites to cities have been in places with convenient access to water, food, farmland, energy, and attractive views. It was no accident that cities of the eastern United States, including Trenton, Philadelphia, Georgetown, Raleigh, and Durham were developed along the fall line between the rock of the Piedmont and sediments of the coastal plain. There was water and power available there and arable land. Only recently, when people gained easy access to the cheap and portable energy of fossil fuels, did that dependence on local resources become obscured. Now as the need to make an abrupt shift in the sources and use of energy has emerged, an intensified interest in the environment has also emerged. In this second decade of the twenty-first century, the global commons are being eroded at an unprecedented rate, and the climates of the earth are being modified to the point of presenting all life globally with the potential for a shock no less threatening and devastating than the 2010 earthquake was to an already-impoverished Haiti.

* * *

A group of scientists in Woods Hole and Acre, Brazil, in 2012 addressed the challenge of restoration from the depths of impoverishment that Lester Brown has described so well. Haiti was of course a major topic after the earthquake. The question was how to proceed with reestablishing a viable political-economic system capable of supporting ten million people on an island of finite potential. There was no clear opinion among the participants that ten million could be supported using existing intrinsic resources at any point, and all had to conclude that external resources would have to be supplied. Then the question turned to what comes first: reestablishing a government, an economic system, or functional landscape. The group's almost-unanimous initial response was a government and then an economic system. But a government, too, requires a place to stand, a water supply, an income, and the respect of the public. The problem appeared at first as the classic chicken-and-egg discussion. In the end it became clear that all must occur at once: a systematic approach toward rebuilding a biophysically sound landscape capable of supporting a modern society, a government to design and implement it, and an economic system to support all. To do it in Haiti would almost certainly require massive outside help to enable the restoration of drainage basins, water supplies, productive farmlands, coastal fish and fisheries, and support for small, local businesses and industries. As the discussion continued, it became apparent that the restoration of Haiti engages many of the same questions and responses appropriate to the global dilemma of the population-pushed world in an age of climatic crisis and imperative energy reformation. Are there lessons here? On reflection, I thought so.

Who will envision and implement such a new order of life in Haiti after two centuries of failure? Why does it not emerge now with so many models to work from? What has been missing on this small island that has allowed thugs and thuggery as well as wanton environmental destruction to prevail over centuries to generate the current abysmal disaster? I do not pretend to have the answers, but the issues might be dissected for further insights. Any move toward restoration of what might be considered a functional modern civilization in Haiti or elsewhere must offer each participant, each individual, something of value in life substantially and immediately. In an increasingly densely occupied world, there are two routes.

The common route is simply an increased intensity of competition to protect or advance one's interests in survival. Competitive thuggery assures the destruction of all commonly held resources and probably most privately held resources as well. It matters little whether the competition is individual and personal or corporate; the product is the same: the systematic destruction of common property resources. Haiti is the most extreme example in the Western Hemisphere, although there are abundant other, less extreme examples even within the United States, where well-recognized laws have been flouted, as discussed above.

The alternative starts with the recognition that certain essential requirements for life are held in common, and can be met equitably and reliably at least cost by joint agreement and joint action. The thought starts with simple rules—manners of long standing such as the golden rule virtually universally recognized as the foundation of reason in behavior and law. It is then a small step to recognizing an array of civil rights and universal needs such as a reliable and safe water supply that can be made available to all by restoring drainage basins, starting small and moving up with success and time. The problem is that thuggery persists and is contagious, as we observe regularly in international and even national affairs. Stifling it requires persistence and ingenuity and a widely shared commitment that it will not prevail.

* * *

At a larger scale, avoiding a global collapse requires a revision of purpose to the recognition that human rights to clean air and water and a place to live are vulnerable to personal and corporate greed. Developing new rules defining essential common interests and protecting them is both necessary and possible. But they will emerge only if cultured and defined, celebrated and respected, by an aware public willing to defend essential birthrights as unassailable civil rights. Those civil rights, protected, assure the physical, chemical, and biotic integrity of the biosphere, the sine qua non of the continued success of civilization. They must be recovered from, or replace, the corporate greed that has stolen them, and given us all a world poisoned into instability and sliding into chaos. The enormity of the threats is giving rise to new perspectives on science, public interests, morality, and human rights. They reach as well into corporate power and purpose, which must extend not simply to profits but also to public service in the world.

Is it too much to ask that in a shrinking world all business and commerce operate within rules consistent with restoring and protecting the physical, chemical, and biotic integrity of the earthly habitat and with respect for the civil rights of all members of present and future generations?[34]

The rule of a minority, as a permanent arrangement, is wholly inadmissible; so that, rejecting the majority principle, anarchy or despotism in some form is *all that is left.*

—Abraham Lincoln, first inaugural address

Fifteen years after the end of the largest and worst war the world could then conceive, the world's two major economies were energetically engaged in developing an even greater horror. Their energy and wealth were being poured into building two competitive arrays of intercontinental ballistic nuclear missiles, each with several independently targetable warheads, enough to incinerate every college and university and to salute similarly with individual distinction most of the schools in each of those great nations. Not that the writhing residual mass of humanity left after the Big Bang would be much interested in education. It was just that there was no limit on the number of weapons, making them was good business, and boasting about them internationally seemed to generate respect—when it should have generated fear and challenges to the sanity of political leadership in both nations. But the competition kept the money flowing into military interests that might otherwise have had to scramble for recognition. Fortunately, the conflagration never occurred. Or, at least, it has not occurred yet despite the existence of the weapons, many provocations and threats and as with any such risky business, near disasters.[1]

The continued military confrontation defined governmental purpose and pushed civil rights, environmental concerns, and the basic science of health and welfare to the side as secondary matters, at least momentarily. But that unfortunate dismissal could not last. By the 1960s the world's ballooning population and a host of environmental issues, including early warnings about heat-trapping gases and climate were already changing the world.

The scientific community, some of it at least, saw the potential for climatic disruption for what it is: a no less definitive end to this civilization than the nuclear war, so long cultivated.

Late in the nineteenth century a prominent Swedish chemist, Svante Arrhenius (1859–1925), had applied his extraordinary talents to appraising the roles of carbon dioxide and water vapor in affecting the temperature of the earth. He had shown as early as 1896 that burning fossil fuels would warm the earth. His calculations suggested that a doubling from the approximately 280 ppm carbon dioxide of his era to 560 ppm would raise the temperature five to six degrees Celsius (nine to eleven degrees Fahrenheit). (Current perspectives suggest that feedback effects triggered by such a warming could easily push the temperature far higher—a catastrophic change in a short time.) With half a century preoccupied by two world wars and a crushing economic depression, the 1896 insights of Arrhenius lay substantially dormant until Revelle of the Scripps Institution of Oceanography, drew Keeling into what proved to be a full career documenting the accumulation of carbon dioxide in air. The discussion opened vigorously as data began to accumulate. As we saw in chapter 4, Keeling found that the carbon dioxide content of the atmosphere had risen substantially above the approximately 280 ppm of Arrhenius's time. By 1958 it had reached about 315 ppm. By the early 1960s it had become clear that the carbon dioxide concentration in the atmosphere was rising annually at 1 ppm or more. That change constituted a significant change in the chemistry and physics of the entire atmospheric envelope of the earth. The issue was important enough that Revelle, who had led a sub-committee of the President's Science Advisory Committee, called the report to the attention of President Lyndon Johnson, who took an interest. Johnson, with the help of Stewart Udall, then Secretary of the Interior, wrote it into his February 8, 1965, Special Message to Congress, the first official mention of the issue by any president, now more than fifty years ago.[2] Scientific interest moved on over the ensuing decade but political interest languished.

The election of Jimmy Carter in 1976 brought a great sigh of relief in science and conservation circles along with a surge of optimism that environmental issues would flow to the top of the agenda. And they did. The most visible early indication was the installation of solar hot water panels on the White House. More important, however, was a special focus on the climate issue in the recently created Council on Environmental Quality in the Executive Office. As chairman of the Council, Carter appointed Speth, a founder in 1970 with colleagues from Yale and others from New York of a muscular conservation law agency, the Natural Resources Defense Council (NRDC).

Speth made stopping the climatic disruption a major issue of the Council on Environmental Quality Among other initiatives sought in 1979 a statement from the scientific community that could be used to draw further attention to climatic disruption. Somewhat reluctantly, for much had been written on that topic, and ignored, over the previous decade, I agreed to draft a statement. Keeling, by then twenty years into his measurements, Revelle, and Gordon MacDonald, a well-known geochemist and member of the National Academy of Sciences—all friends and distinguished scientists—joined without a moment's hesitation. We wrote a comprehensive statement that said in part:

> Man is setting in motion a series of events that seem certain to cause a significant warming of world climates over the next decades unless mitigating steps are taken immediately. The cause is the accumulation of CO_2 and other heat-absorbing gases in the atmosphere. The result is ... a warming that will probably be conspicuous in the next twenty years. If the trend is allowed to continue, climatic zones will shift, and agriculture will be displaced. Such a series of changes would have far reaching implications for human welfare in an ever more crowded world, would threaten the stability of food supplies, and would present a further set of intractable problems to organized societies."[3]

The statement was widely distributed and proved sufficiently provocative to prompt congressional hearings. It also served as background for a comprehensive, detailed statement on the Council's vision of its mission commissioned by Speth, *The Global 2000 Report*, published in 1980. The report anticipated the environmental status of the nation if we followed a wise course in management to a date that seemed then far, far off.

The presidential election campaign of 1980 pitting Carter against Ronald Reagan was propelled by a quite different issue: continuing confrontation with Iran over its internment of the US embassy staff. The issue was compounded most unfortunately by a botched military attempt at their rescue. Carter appeared, momentarily at least, as ineffectual. The circumstance was used in advancing Reagan's popularity. Carter had in fact negotiated the release of the hostages, but the release was delayed, perhaps by Reagan interests, until after the election when Reagan was in the White House in early 1981.

Reagan's inauguration brought an abrupt change in environmental policies and perspective. The new administration soon removed the White House solar panels Carter had installed and made sure that *The Global 2000 Report* was not widely distributed. Those were but the initial celebratory steps in advertising Republican hostility to renewable energy, opposition to advances in protecting common property resources from destructive private use, refusal to acknowledge the looming climate crisis, and hostility to responsible management of public lands and resources.[4] Even now, more than thirty years later, the entire world continues to suffer from the Reagan administration's reversal of perspective as to governmental purpose as well as its reversal of federal policies on energy and environmental protection. In failing to start the transition from fossil fuels to renewable energy, the nation lost a magnificent opportunity in economic, environmental, and political leadership.

While the national leadership of the United States has vacillated over decades, either unwilling to take any major action, or, willing, unable to find sufficient support in a reluctant Congress, various nongovernmental or state agencies have been able to move forward with comprehensive programs for reducing reliance on fossil fuels. The northeastern states, for instance, developed the Regional Greenhouse Gas Initiative (RGGI), later imitated by the State of California. It is in effect a "cap-and-trade" system for reducing greenhouse gases under which power companies are required to produce a certain fraction of their power from renewable sources. Unable to provide their own renewable energy on the schedule established, they can purchase emission rights from others who are producing renewable energy at or above needs or requirements. Potential suppliers of renewable power who could sell "rights" include homeowners who install wind turbines or solar panels. These programs have been highly successful and have been encouraged by tax subsidies in the federal tax system. Such innovations have been imaginatively developed and encouraged by nonprofit agencies such as the Environmental Defense Fund, the Natural Resources Defense Council (NRDC), the Rocky Mountain Institute (RMI), Common Cause, and many others, all of whom have contributed to a vigorous public discussion over years. The power of these nonprofits has been a breath of fresh air in government in a time when fresh air was much needed. My personal experience has been most intensively with the NRDC, a prime example of the significance of nongovernmental agencies in keeping a functional governmental system in rapidly changing times.

By 1982, the NRDC, already in its thirteenth year, was one of the most powerful nongovernmental conservation agencies in the world. Initially the organization had taken as its mission forcing the government to obey its own laws. The National Environmental Policy Act (NEPA) was signed into law on January 1, 1970, the year the NRDC was founded. The act established the novel requirement that new projects would be required to have a comprehensive Environmental Impact Statement (EIS). This review offered a new set of insights into major projects and opened a basis for questioning core assumptions as well as details of design. The requirement for an EIS advanced the needs for data, frequently more research, and scientific insights into the environmental effects of every new project that required governmental participation or approval. The NRDC often found itself defining and defending the public interest and welfare in the broadest context. The courts offered a powerful tool, but so did the dogged presence of talented and experienced lawyers expert on the topic at hand. Over years the NRDC and others accumulated more experience in the law and on critical environmental topics than existed in any governmental agency, business, or industry. And it had become clear that as society expanded, more and more of the business of government, including economic and political developments, had substantial environmental components that required detailed review. The Pebble Mine, a giant minerals mine in Alaska that would drain into Bristol Bay, is a recent example. The EIS reported the size of the mine as well as the effects it would have on land, streams, salmon stocks, and Bristol Bay itself. The public is finding the project simply too much, too large, and too damaging for the entire region. Public pressure brought mainly by the NRDC on government and on the mining corporations appears to be on the verge of forcing the abandonment of a seriously destructive project, which this writing (2015), seems on the verge of collapse.

With the advent of the Reagan years, Speth, widely respected but personally scorned by the new political regime, had with others at the NRDC moved to create a nongovernmental institution in Washington, DC, to explore emerging environmental policies and needs—a project previously conceived and long in gestation. Speth enlisted support from his admirers in the environmental community and in the world of private foundations.

He drew in the talents of Jessica Tuchman Mathews as a close colleague, and with major help from the MacArthur Foundation, founded the World Resources Institute (WRI) and became its first president.

The World Resources Institute took on as its institutional business the structure of government and the principles involved in accommodating the march of environmental intensification. The climatic disruption was, and remains, at the forefront of the Institute's activities, although the confrontation of corporate and industrial interests, the public interest, and human welfare were seen as its special province.

Establishing a new, well-financed, and intellectually and politically progressive environmental institution in Washington was a major advance, and its emergence was widely noticed. I was pleased to have been involved as an active member of its board and vigorous supporter of its distinguished staff. The World Resources Institute's presence established powerful advocates in the public interest who had access to government at all levels including Congress.

* * *

A signal event in US public recognition of the dangers of climatic disruption was a set of hearings before Congress in early summer 1988. On June 23, six scientists, I among them, summarized scientific perspectives on climatic disruption for the Senate's Committee on Energy and Natural Resources in the first of a series of sessions led by Tim Wirth, then a senator from Colorado.[5] The Senate hearings were followed four days later by Oversight Hearings before the Subcommittee on Interior and Insular Affairs of the House of Representatives.[6] The Senate testimony, while reporting on a broad consensus among experienced scientists, carried for each of us a highly personal element. It reflected intense exasperation at having over the preceding decades defined a serious challenge to human welfare only to be virtually ignored. Underneath the testimony were cries of pain and concern, even terror, over what could happen to humanity if action were not soon taken to reverse the trends in the composition of the atmosphere, so clearly the product of expanding use of fossil fuels.

Testimony of that day on climatic change was noticed by the news media as never previously. It was the day when James Hansen, a government employee of the National Aeronautics and Space Administration as head of the Goddard Space Science Institute at Columbia University, testified that the Reagan administration had attempted to suppress his testimony, but that he had decided to testify anyway to bring to public attention the evidence that the warming of the earth was proceeding and was then measurable. He had led the effort in figuring out how to measure the temperature of the earth systematically and credibly, and was now bold enough to report it to the Congress despite opposition from his employers. There was little news in the fact of the warming. But there was news in the statement that the administration sought to suppress the evidence. That fact made the event important, and in the eyes of many, the date stands indelibly as the occasion on which the first evidence was reported of global changes in climate. It was precisely contrary to the intent of the administration, of course. Subsequent administrations, Republican and Democratic, learned little or nothing from either the testimony or the Reagan fiasco in attempting to suppress it. No action emerged and the fossil fueled age was prolonged augmenting the flow of profits to the giant industries engaged in fueling it with oil, and coal and gas.

What is striking now, more than a quarter century later, about the 1988 Senate testimony, including my own of that day defining the implications of a systematic warming of the earth, is not only that it was correct in its detail at the time but also that with few alterations it might be offered today as evidence in favor of governmental steps in amelioration, still not taken. In view of the developments of subsequent years, especially the great climatic events of global consequence of 2008–2013, the increasing rates of glacial melting, the expansion of arid zones, the greater frequency of severely damaging storms, and floods that in some parts of the world have devastated agriculture, our predictions in 1988 of likely occurrences have been borne out and reported in thousands of news articles.

We knew a lot then. We knew the range of consequences of sending more and more greenhouse gases into the atmosphere and the probable costs, economic, political, and human. There was ample information at the time to start corrective action. We know much more now. And we have not only clear evidence of the changes in temperature predicted but also the costs of effects in our own time. We have had to learn to adjust to a devastating storm-driven flood in New Orleans, deepening drought in the US Southwest and West, floods in the US East, glaciers melting globally, and cities around the world starving for water. We also have Haiti, Somalia, and Pakistan as examples of the political and economic depressions that can include among their causes mounting environmental disasters. We know that global processes of environmental deterioration are moving more rapidly than anticipated, and that the costs are high and rising by the day. And we know that the cures, properly designed, are potentially effective and attractive to society as a whole, financially as well as environmentally. Yet, blinded by inertia, wishful thinking, and the blatant lies of opponents who thrive on the economic power of the current corporate world, we stagger on as a nation, helping to lead the world into the chaos of three-sided collapse—economic, political, and environmental.

* * *

By the closing decades of the twentieth century, major scientific groups and other knowledgeable citizens around the world had come to recognize the potentially disastrous global climatic disruption under way as well as the necessity for consolidated international action to prevent it. The UN Environment Programme was planning a major international meeting around environmental topics for 1992 in Rio de Janeiro, and it was important to enter those meetings with a strategy for deflecting the climatic disruption. My colleagues and I at the Woods Hole Research Center, long involved in the issues, realized that there should be not merely scientific discussions but also international efforts to correct the trends. What should we do?

At that moment an already-distinguished young lawyer from India with a background in international environmental law appeared on our doorstep. Kilaparti Ramakrishna was seeking a temporary position while his wife completed a graduate degree in Boston. His PhD in India had been on the UN system, and he had held a postdoctoral position in the Harvard Law School and, for one year, at the Woods Hole Oceanographic Institution. I thought it a wonderful opportunity, and suggested that he join our staff at the Woods Hole Research Center and draft a model treaty over the course of the academic year. Ramakrishna joined our still-small institution and over some months outlined, not a treaty, but a route to in the UN system. His plan was to engage the UN General Assembly and encourage them to establish a committee to write the treaty. All should be done in time for the treaty to be considered at the international environmental summit meeting in Rio in 1992.

Ramakrishna was well acquainted with the UN hierarchy. At my insistence, he arranged for a review of the proposal by key officials including the then permanent representative from the United Kingdom, Sir Crispen

Tickell, who had written a book on the seriousness of the climatic threat. We all met in my living room in Woods Hole in 1989 to discuss the possibility. Tickell and his colleagues proved receptive, and returned to New York to report to the Secretary General and the General Assembly. A committee was established and was charged with preparing the draft treaty. Ramakrishna's interest in chairing the committee was recognized, but funds were short because the United States at that time, as a result of the shameful opposition of a single senator from North Carolina, Jesse Helms, was not paying its UN dues. Michael Zammit Cutiar from Malta was instead named chair, but he immediately sought Ramakrishna's aid as an adviser familiar with details of the problem. The treaty, the Framework Convention on Climate Change (FCCC), was drafted, presented at the Rio meetings in 1992, and to the great credit of the negotiators and the national participants, signed by virtually all nations—a remarkable and unusual success. It was ratified in the following months almost universally around the globe. Ratification included the United States, and it is today the law of the land.

In ratifying the document, nations agreed to stabilize the heat-trapping gas content of the atmosphere at levels that will protect human interests and nature. The level was not defined, but it became clear from the chain of effects predicted and already occurring that it should be below the 350–370 ppm of that time. The treaty was a statement of intent, not a design for action. Its implementation was left to a future agreement, but all nations were free to act independently to accommodate the objectives already established. As in most international agreements involving many other nations, US leadership was obviously crucial.

The next step came slowly. Five years later in 1997 a Conference of the Parties (COP) to the FCCC attempted to develop the first clear commitment among nations to achieve the objectives already agreed to under the treaty. A draft protocol was prepared, negotiated, and signed on December 11, 1997, in Kyoto. It, too, was a remarkable achievement, that recognizing the large role that the developed nations have played in building the current problem and provided flexibility by allowing them to trade carbon dioxide emission rights. The protocol established binding agreements among the industrial nations to reduce their emissions over a finite period. It also advanced a "Clean Development Mechanism," which favored the interests of developing nations in benefiting from use of fossil fuels. Subsequent meetings extended the agreement and commitments.

The Kyoto Protocol to the FCCC was the product of extraordinary efforts by the negotiators, who made accommodations to the needs of less developed nations and provided flexibility in the potential responses of nations. But the US Congress did not ratify the protocol—an essential step in gaining international implementation. That failure has been a tragedy, compounded through successive failures of Conferences of the Parties to arrive at any universally acceptable agreement on reducing emissions, even as the climatic disaster has laid itself before the world in ever more terrifying detail.

By December 2013, we in the United States could look back on a two decadal succession of summers of record temperatures in the Northern Hemisphere, record droughts, and record floods, and we could appraise the growing global climatic chaos as well as its cost in lives, accumulating misery, and economic disaster. All were ignored or actively denied as climatic issues of significance by a powerful Republican faction, particularly in the US House of Representatives, dead set on crippling government action in every way possible.

The focus was and remains on the United States as the key player on the global stage. The Wolin observation of inverted totalitarianism, with profit-seeking corporations writing the laws to favor themselves, gained further support as corporate political and economic power accumulated, and public welfare dropped below corporate welfare as a governmental purpose (see chapter 5). Such are the fruits of the current run of neoconservatism in which the fundamental purposes of government are seen as support for a corporate structure that enhances corporate profits and promises that largesse will trickle down to the populace in the form of jobs and other benefits. Corporate interests extend extragovernmentally to international relations. Despite national efforts to limit oil development in Arctic waters and economic ties to Russia to discourage Vladimir Putin's 2014 adventures in Ukraine, Exxon Corporation expanded its extensive joint efforts with Russian oil interests. Exxon's responsibilities are regularly candidly defended in relation to stockholders' profits, not to loyalty to national objectives or human welfare.[7]

National elections in the United States have of course also made a major contribution to the current predicament. The appointment of George W. Bush to the presidency in the 2000 election by the US Supreme Court marked, after Reagan, a second unfortunate reversion of presidential leadership to doctrinaire aversion to resolving the climatic crisis. It also assured that the political and economic progress of the previous eight years, limited though it was under Clinton and Gore, would be deflected as well. By the end of eight years of the George W. Bush administration, the nation was engaged in two wars, suffering an expanding financial crisis, and had energy policies that ignored the climatic disruption and were indeed making it rapidly worse, notwithstanding the objections of a growing cadre of citizens and scholars.

But the facts of the world remained, and experience accumulated despite presidential biases.

Pakistan had vast areas of its best agricultural land inundated in the summer of 2010 for weeks by floods and removed from production for a second season; its economic system, ever fragile, was beset and paralyzed; its government, weak and tenuous in its hold on authority, let alone the public welfare, was also beset and ineffectual in rising to the new challenges.

The grain crops of Ukraine and Kazakhstan in 2011 were afflicted by drought to the point where Russia, fearful of a food shortage, proscribed foreign sales. Prices rose globally in response, just one of many effects that environmental stress can have on governments and the global economic system. North America suffered a record continent-wide drought. The *New York Times* carried a front-page article reporting not only the continued drought in Kansas but also the depletion of the once-abundant groundwater used to irrigate bumper harvests of corn:

> Forty-nine years ago Ashley Yost's grandfather sank a well deep into a half-mile square of rich Kansas farmland. He struck an artery of water so prodigious that he could pump 1600 gallons to the surface every minute.
>
> Last year, Mr. Yost was coaxing just 300 gallons ... pumping up sand ... to do it....
>
> "That's prime land," he said.... "I've raise 2194 bushels of corn an acre there.... Now," he said, "It's over."[8]

Not only has drought had profound effects in Kansas. It has also affected much larger regions of once-arable land in the US Southwest and Mexico. Part of the well-advertised pressure on the US southern frontier is from Mexican families forced off the land by a drying climate and seeking a new life. Elsewhere, island dwellers around the world were finding themselves beleaguered by higher tides, larger storms, and anticipation of a rapidly rising sea's claiming more and more of their living space.

Despite the monumental effort at Kyoto in 1997 and some exemplary actions by some individual nations, no proposal or emergent biophysical

threat has yet been sufficient to drive the nations collectively to any significant step toward abandoning fossil fuels, and in that way, setting the
world on a course to preserve the essential qualities of the biosphere. The
failure of the nations to join in effective action must be laid primarily at
the feet of the United States because the US Senate expressed opposition
to any binding agreement such as that of Kyoto. The opposition has continued through successive administrations, fed by narrow venal interests
supplied with virtually unlimited sums of money to pour into influencing
Congress and the public to avoid any step to inhibit extraction and the use
of fossil fuels.[9] Climatic effects have accelerated through all those years,
and yet have been ignored or even denied systematically by politicians,
many of whom have been funded directly by coal and petroleum interests.
Distortions have run wild, and to some became a fanciful belief system that
denies the reality of biophysical facts and natural laws. Most notoriously,
James Inhofe, a Republican senator from Oklahoma, twice appointed chair
of the Senate Committee on Environment and Public Works, in 2012 published a lengthy statement denying the reality of the climatic disruption
and the science behind it, claiming in part:

> Thus far no one has seriously demonstrated any scientific proof that increased
> global temperatures would lead to the catastrophes predicted by alarmists. In fact,
> it appears that just the opposite is true: that increases in global temperatures may
> have a beneficial effect on how we live our lives.[10]

Such sentiments are popular among those who have heavy financial
investments in any aspect of the fossil fuel business. They find allies in
commerce and build tax-exempt (501[c][3]) institutions such as the American Legislative Exchange Council, which emerged in the early 1970s to
develop the right-wing agenda at the state and national levels. Their purpose is not to enhance the public's welfare or sound government but to
bend legislation and legislative activity to favor corporate interests. To take
an example, fossil fuel interests and their compatriots launched a conspicuously scurrilous campaign to challenge the veracity and accuracy of data
and analyses produced by a highly respected scientist at Pennsylvania State
University, Michael Mann. But time after time the quality and accuracy of
his work had been upheld, and greatly praised, by independent reviewers
of his scholarly papers.[11]

For some, especially those with tentacles firmly fixed in the fossil fuel
industries, opposition to action on the climatic disaster remains adamant.[12]

For others, however, there was in reality only one possibility: turning back the oil spigot (and coal and gas) and making every effort to return to a stable climatic system more akin to that of 1900 than that of the present. In today's world, an appropriate transition will require immediate accommodation of changes already existing, and those certain to come, as well as abruptly reducing emissions and reconstituting every conceivable sink for carbon.

As effects of the climatic disruption have accumulated, the awkward necessity of some grudging acceptance and accommodation of existing change has become obvious even to those who have never taken seriously the implications for our collective future but assume we will somehow "muddle through." But "adaptation" taken as an end in itself is no solution.

THE ADAPTATION MYTH:
AN ATTRACTIVE CONCEIT
THREATENS ALL

What's the use of having developed a science well enough to make predictions if, in the end, all we're willing to do is stand around and wait for them to come true?

—Sherwood Rowland, quoted by Elizabeth Kolbert, *New Yorker*, April 14, 2014

As the nineteenth Conference of the Parties to the Framework Convention on Climate Change gathered in Warsaw in mid-November 2013, the island nation of the Philippines struggled to recover from the giant storm, Haiyan, of November 7–8. The storm's extraordinary dimensions, a tidal surge of twenty-five feet accompanied by winds of two hundred miles per hour that caused thousands of deaths across an island archipelago, might seem to have put into urgent terms the mission of this gathering of delegates from potentially more than 190 nations. But as with its eighteen predecessors, inertia was overwhelming and movement was imperceptible.

At the same moment oceanographers were planning details of what they considered urgent new studies of the circulation of the waters of the North Atlantic. Their focus was on the quite-realistic possibility that the warming of surface waters is increasing the stratification at the surface and slowing the mixing of surface water into the depths. The consequences of such a change might affect the rate of warming appreciably.[1] The transfer of carbon dioxide from the atmosphere into the deeper, colder waters would be reduced by both the reduction in mixing and the reduction in absorptive capacity for carbon dioxide of the warmer surface waters. At the same time warmer waters vaporize more water, thereby adding to the energy in the atmosphere (energy of vaporization) that drives storms such as Haiyan.[2] These are highly threatening transformations whose only cure is a reduction in the energy available in the atmosphere. The first step in that direction involves reducing the atmospheric burden of carbon dioxide to levels far below the 400 ppm of today. Again, experience is showing that safety lies in a reduction in atmospheric carbon dioxide over this century to approximately the level of 1900, 280–300 ppm by volume in air. Even at 300 ppm, the concentration of carbon dioxide would be higher by about 20 ppm than it has been in the previous eight hundred thousand years.[3] Unfortunately, no governmental or other program, nationally or internationally, has established such an objective, and the global upward trend in atmospheric continues virtually unchecked despite the regular COP meetings. Meanwhile, the problems accumulate.

Popular discussion claims that a rise of two degrees Celsius in the average temperature of the earth above the temperature prevailing at the beginning of the twentieth century would be acceptable. That proposition has no basis in science; quite the opposite. It was an attractive but seriously misleading "compromise" among economic and political interests, accepted by some scientists but recognized as a compromise. It takes little expertise in science to observe that the global warming already experienced, an average of approximately 0.8 degrees C, has become seriously burdensome and is triggering further changes of great consequence.[4] One of the most disturbing effects is the warming of the boreal and Arctic zones, where the temperature increase, discussed earlier, exceeds the average for the earth as a whole by two or more times. Allowing further warming at 0.15–0.2 degrees Celsius per decade would be the height of folly.[5]

The segment of the corporate world that is heavily engaged in producing and using fossil fuels finds any justification for delay in restricting greenhouse gas emissions attractive, and the myth of a two-degree safety factor is attractive. In this make-believe realm, repetition creates reality and words take on new meanings. The insistence that the two-degree myth has support in science has become widely used as a license to delay any action.[6] Action—a tax on fossil fuels, for instance—is displaced by talk about how we can and must "adapt" to the changes already evident as well as the further changes that we cannot avoid. That we must adapt to changes already entrained is the soul of reason. Of course we must avoid damage from larger storms and higher tides. But as an excuse for inaction in confronting the urgent need to reduce the burden of greenhouse gases, it is madness. Nevertheless, adaptation, accommodation to the accumulating disruptions, becomes both reality and policy of powerful well-financed interests in government and industry.

Utterances about adaptation are usually accompanied by the assertion that mitigation must occur where possible. "Mitigation" refers to reducing the causes of the disruption, which is much more difficult and too conveniently left for some ever-receding future moment. Repetition soon transforms adaptation to a policy of muddling through with at best only a trifling change in behavior—the far preferable policy for the largest and richest corporations in the world. Thus words become political tools. Adaptation soon and mitigation sometime later become policy, political and economic shelter for interests that intend to continue putting profits ahead of the public welfare for as long as possible. Meanwhile, the US policy of year-by-year accommodation fails to address the underlying issues. John Holdren, the Obama administration's senior advisor on science and technology, set forth a practical approach for government in 2010:

> Adaptation alone won't work because adaptation gets more costly, more difficult and less effective as the size of the climatic changes to which you're trying to adjust get bigger. And therefore what we need is enough mitigation to avoid unmanageable climate change and enough adaptation to manage unavoidable climate change.[7]

Holdren caught the reality of the political circumstance, but slipped over the seriousness of the emergency and the irreversibility of the daily further slide into global environmental chaos. He left no question as to his personal understanding of the issues. His words covered but did not justify a national policy of accommodation and adaptation with mitigation when convenient.

While celebrating higher standards for the gasoline mileage of automobiles and subsidies for renewable energy, the Obama administration justified on economic and political grounds further development of new sources of fossil fuels with the prospect of achieving national "energy independence." If independence could be based on a massive infusion of renewable energy and a parallel drop in both production and use of fossil fuels, the current administration and the United States as a nation could stand before the world as a leader without equivocation. Instead, we continued to offer the largest and wealthiest corporations in the world the largest subsidy ever advanced: the right to poison the biosphere, the entire human environment, indefinitely for private profit. Government after government around the world has also succumbed, paralyzed, trapped in an open-ended mission of adapting to current climatic changes, offering words favoring mitigation of the problem, but in fact, by delaying or offering only token changes, speeding to catastrophe.

So it was until May 2014, when in the United States, President Barack Obama announced an administrative action under the EPA's authority to phase out coal-fired power plants over the next decade.[8] His decision involves the single-largest source of carbon emissions in the nation, or about one-third of the total. Although it was late in the development of the problem, the US initiative constituted the largest step that any nation in the world has taken to date toward reducing carbon emissions. It will be opposed vigorously by the full spectrum of fossil fuel interests, which will also push gas as the alternative. While the transition away from coal seems an obvious first move, it has been long in coming, and despite its importance, is a small part of what must be done to gain control of the global climate and the approximately eleven billion tons of carbon released annually into the atmosphere worldwide. Procrastination remains the policy, and the costs accumulate.

Delay is a commitment to continued emissions, despite the adoption of some laudable practices around the world. For the moment, there will be both a continued buildup of effects and continued attempts at repair. Delay ignores the urgency of the situation and the inexorable nature of the consequences if little is done. It ignores the droughts and crop failures, the multiple hegiras of Africa and the Americas, melting of the soils of the Arctic and potential flood of carbon dioxide and methane that may emerge, the fires and insect pests sweeping the boreal forests of the Americas and Asia, paralysis of entire regions as summer temperatures soar to lethal levels and raise mortality rates, the frequency of disastrous storms of unprecedented ferocity, and coastal flooding and rapid erosion. And it discounts the future as glaciers melt and sources of water for tens of millions shrink, possibly disappear, as sea level rises, first slowly, then rapidly tens of feet, ultimately more than two hundred feet unless we are soon successful in reversing the trends of climate. Since the environmental revolution of the 1970s, we are already more than forty years into the Age of Climatic Disruption and the devastating trends continue. Worse, some, including some scientists who should be offering better advice, continue to advance the proposition that a two-degree warming would be safe and acceptable.[9]

The political schism of the early years of the new millennium in the United States has pitted conservative ideology against virtually any application of environmental science in public affairs. To the extent that there is a philosophical core to the ideology, it is that old warhorse, an unfettered free market's capacity for self-regulation and accommodation (or adaptation), that will eventually ride to rescue the public interest and welfare. The perspective is currently popular among corporate interests despite the evidence that they, too, benefit from sure-handed governmental regulation and stability.

The business-as-usual ideology appeals to virtually all elements in commerce, requires no governmental action, and enables continuing high profits from lax regulation of industries that exploit commonly held resources. The fossil fuel interests have justified spending many millions in denying the reality of the climatic disruption with statements suggesting that reducing emissions of heat-trapping gases would be, in the words of Rex Tillerson of ExxonMobil, "devastating to economies, societies and people's health and well-being around the world."[10] The oil business has been highly profitable for many decades, and there are sufficient additional energy reserves in the ground to support an unfettered industry for years to come. The new techniques surrounding fracking only open new reserves. To a petroleum executive like Tillerson, any policy that delays or deflects restrictions on development of fossil fuels and their use is attractive, although it has been apparent for forty years that the public welfare would be best served by policies that are the polar opposite.

Tillerson is not alone. One of the Koch Brothers, William, has, for example, been instrumental in delaying for twelve years James Gordon's development of the first US offshore wind farm. The project, if completed, would produce electric power equivalent to three-fourths of the base load of Cape Cod, Nantucket, and Martha's Vineyard. It would displace power from a giant coal plant on the southern coast of Massachusetts and nuclear plant in Plymouth, south of Boston. Koch was quoted directly as pursuing two strategies to defeat the project: "One is to just delay, delay, delay, which we're doing and hopefully we can win some of these bureaucrats over.... The other way is to elect politicians who understand how foolhardy alternative energy is." Koch has financed this and similar efforts over more than a decade with millions of dollars to delay renewable energy projects as long as possible, and protect his interests in various aspects of the oil, coal, and gas economy.[11]

Koch and Tillerson, however, are but two of the most prominent among those in the economic elite who argue either that concern about climate change is overblown or that, even if the climate is changing, its effects can be readily accommodated as they occur and there is no need to worry about using fossil fuels now or in the future.

Meanwhile the damage accumulates and the public suffers, both financially and, worse, from an environment that is progressively impoverished and difficult to live in. Flexibility in building schools and other public facilities is lost as money is drawn off into restoration of storm damage as well as defenses against further damage and disease. Resources for accommodation of the disruptions and repair are limited, and many repairs never occur. So it has been with New York and New Jersey in the aftermath of Sandy, and New Orleans and the Gulf Coast after Katrina. Much of the devastation is beyond financial reach, requiring abandonment of a region and reconstruction on a new landscape. The human birthright to a decent life erodes rapidly, and many governments, despite a nominal interest in protecting the public, are speeding the erosion. We adapt to the new reality, forced by circumstance to accept impoverishment. Adaptation is attractive, even contagious, when it refers simply to accommodating changed circumstance, with no attention to underlying causes, the magnitude of the costs, or what might come next. Adapting becomes long-term policy: the assumption that we can all adjust to troubles, continue on the same course indefinitely, and hope always for a better world somewhere just over the horizon.

* * *

The most insidious error in these assumptions is to overlook the feedback problem. The effects of the warming tend strongly to enhance the warming. As we saw with the oceanographers' concerns about the slowing of the oceanic circulation, as the surface water warms, the accumulation of carbon dioxide in the atmosphere is amplified and the energy available through evaporation for driving atmospheric circulation (and its storms) increases as well. In another example, the summer reduction in the ice cover of the Arctic Ocean increases the area of open water, a dark surface that absorbs additional solar energy. The warming of the surface water further increases the energy entering the high-latitude atmosphere as water vapor. The effects on climates in the Northern Hemisphere are simply unknown, but the assumption that there will be no effect is naive in the extreme. The extent of the thaw is accelerating, and the entire Arctic Ocean is expected to be open in summer within less than two decades.[12]

More generally, the oceans have long been recognized as having a major influence on the composition of the atmosphere. The primary influence is the annual exchange of carbon dioxide between the surface waters, fully two-thirds of the earth's surface, and atmosphere. An increase in the amount of carbon dioxide in the atmosphere favors a net diffusion into the oceans. Higher oceanic temperatures push the flow of diffusion in the other direction and currently slow the net transfer from the atmosphere to the oceans. The more carbon dioxide that accumulates in the ocean, the more acidic it becomes, and the more difficult it becomes for organisms to make and retain carbonate shells, which itself has a devastating effect on ocean life. In the short run of years, the net flow of carbon currently is from the atmosphere to the oceans, globally about two billion tons of carbon annually. The vast store of carbon in the oceans will be released slowly if the atmospheric burden of carbon dioxide declines in response to efforts to stabilize the atmosphere at a lower concentration. That large oceanic burden makes the reversal of current trends increasingly difficult the longer that current flows prevail and the more carbon we allow to be transferred to the oceans.[13] The most powerful elements in accentuating the trends toward rising temperatures and increased climatic disruption lie in the biotic effects of chronic disturbance that were discussed earlier. The global vegetation, including especially forests because of their large area, is engaged in annual exchanges of carbon with the atmosphere, as explored in chapters 2 and 3. These exchanges vary with seasons and are sufficient to change the carbon dioxide content of the atmosphere by a few percent seasonally. Such variations reflect the balance at any time between the absorption of atmospheric carbon dioxide through the photosynthesis of green plants, and its release through the respiration of all living tissues as well as the decay of organic matter in soils, bogs, and swamps. While photosynthesis is responsive to light, respiration is largely controlled by temperature and moisture. (We take advantage of that fact in protecting food from decay by refrigeration.) Forest trees, developed over decades to a century and more, are weakened, and become vulnerable to disease and insect pests as the climatic regime shifts. A warming of the boreal zone, for instance, already as much as 2 degrees Celsius in some places, increases the demand for water in summer. Trees stressed by the new climatic regime become vulnerable to insect pests such as bark borers. The mortality of trees rises, and their removal of carbon from the atmosphere ceases. As forest trees die and decay, that carbon, too, is returned to the atmosphere as carbon dioxide and methane. Organic matter in soil, exposed to the sun and the new climate, is warmed, and if moisture is available, becomes vulnerable to decay.

In addition to these direct influences on the metabolism of the landscape, the warming climate is making the circumpolar boreal forest, the high-latitude forest of North America and Eurasia, drier and more vulnerable to fires. This is the largest forested area in the world. Those fires, too, release additional carbon into the atmosphere—a further awkward result of continued warming.

The problems of positive feedbacks (those that intensify a process) are even greater in the Arctic, in part because the rate of warming there is greater. The Arctic north also contains large quantities of carbon, especially in the soils and deep peat of the tundra. Soils, long frozen, have trapped the gaseous products of slow and intermittent anaerobic decay over thousands of years. Thawed, the soils release the gases, including methane and carbon dioxide, and release as well large pools of carbon in soils potentially vulnerable to more rapid decay in the warmer climate.[14]

The total carbon held in soils and vegetation globally is variously estimated to be approximately 60 to more than 200 percent of the approximately eight hundred billion tons of carbon currently held in the atmosphere. While there is no proof that the climatic disruption will release a substantial part or all of that carbon, the possibility is sufficiently great and the consequences sufficiently serious that no such risk can be acceptable. A substantial release would mean an uncontrolled and uncontrollable climatic excursion of unpredictable proportions. Adapting to such changes would be a far reach—a dream.[15]

Reasonable analysts, including political leaders, find themselves confronting current changes in global climate and may decide that those changes must simply be accommodated as a matter of policy. That may make some sense in the short term, but it is no substitute for changes necessary in the long term. The "rescue" of New Orleans in the wake of Katrina, for example, has entailed a renewal and extension of levees, better pumps, and more energy applied to mere survival. It has also required at least tacit acceptance that a city, currently one to two feet below sea level, will be expensive to maintain as the sea level continues to rise and the city itself subsides into soft sediments. Accepting this burden entails recognition that the city is not simply accommodating a new stable state of sea and land; it is adjusting to the present and anticipating more adjustment year by year—a continuing and quite possibly accelerating, expensive challenge into an indefinite future. For the moment, the city can adapt to the current

circumstance and recover its stability. For the moment. A longer-term viable future exists for New Orleans only if the sea level can almost immediately be stabilized, which is an unlikely turn for the world as a whole given
current trends and the modesty of efforts to reverse them.

* * *

A core of long-experienced scientists who have some basis for analysis of
the structure and function of the biosphere as a biophysical system, along
with those who follow their work, see the climatic disruption for what it is:
a continuously cascading disaster for this civilization. The formal warnings
are many, advanced over decades and repeated regularly from the highest
levels of our scientific advisory agencies, including the National Academy
of Sciences' National Research Council.[16]

Many others, including economists and politicians as well as scientists
less experienced but interested, express a range of views from disbelief in
the details and basic data to mild concern that the scientists may be correct,
but claim uncertainty and no need now for serious changes in commerce or
society. In the face of uncertainty, little or nothing should be done, many
believe. This latter group includes those who argue that changes are inevitable and humans, aided by technological fixes that are sure to appear in the
nick of time, can adapt to all, so "bring it on!" These latter, of course, include
the Tillersons and Kochs of the world with economic or political interests in
the status quo who have cultivated uncertainty as a basis for inaction.

Unfortunately, as we've seen, warming can feed further warming. Carbon dioxide warms the earth, and the warming produces carbon dioxide.
The feedbacks amplify the effects. Allowed to run its course by ineptitude
or unwillingness to check the trend, the climatic system will follow a devastating course. Pushed to extremes, life will survive. Civilization will not.
The biotic feedbacks are clear, inevitable, anticipated, and defined—and
still potentially controllable, but only if we abandon the production and
use of fossil fuels substantially and immediately with the purpose of reversing the trends in global climate.

The problems of global warming do not all come slowly over years. They
may come suddenly, as many have already witnessed, in unanticipated
storm-driven high tides—storms whose furor is enhanced by the additional
energy entering the atmosphere as water vapor in the energy-dense lower
latitudes.

The most conspicuous and threatening changes associated with the general disruption of climates globally currently are the expansion and intensification of the arid zones of both hemispheres. The continental centers, as long predicted by climatologists, become drier while some sections of the margins grow wetter as the hydrologic cycle intensifies. Now, after the first decade of the new century, there has been a drought on every continent. A particular concern is the diminishing supply of freshwater. Many of the great cities of coastal regions tap mountain streams fed seasonally by snow and ice. San Francisco, Los Angeles, Seattle, and Portland in the United States, and Lima, Peru, and Santiago, Chile, all depend on winter snows. Some, such as Santiago, depend on glaciers.

The summer of 2012 produced the most severe drought across North America since the dust bowl years of the 1930s. That summer simply added another year to the decade-long drought of the American Southwest and Mexico. Along with central Africa, central China is afflicted. Even China's eastern continental cities such as Beijing are looking for water. The latest dream is to move water from the Yangtze in the southwest to the Gobi lands and northeast to Beijing. South America and Australia are both reeling from persistent, devastating dry spells. Adapt to these changes? In Mexico City, adaptation to drought has been to shut off the water supply intermittently, days at a time. At what point does the drought make the city uninhabitable? Move out, but to where?

As continental regions dry out, the migrants of the southern Sahel struggle eastward in search of still-arable land in Somalia or join the thousands seeking to cross the southern European frontier—a sample of hegiras yet to come as chronic disturbances drive larger segments of the earth into biotic and economic impoverishment.[17] The droughts have barely started, but as they develop teeth they will take a monstrous bite from the global food basket; millions will attempt to migrate, and millions will starve.

* * *

Rising sea levels are already a problem in some areas. To put the issue in some perspective, a mere ten thousand years ago as glaciers melted in the Northern Hemisphere, sea level was rising from a low that was somewhat more than three hundred feet below the present level.[18] At the time, a land bridge linked the Asian and North American continents across what is now the Bering Strait.[19] On North America's eastern coast a forest grew on what is now Georges Bank, a hundred miles off the present New England shore. Over those centuries of melting, the average rise of the sea against the land was of the order of three feet per century. Such a change now, even over a century, would be seen as a stunning theft of coastal lands as large segments of Florida, Louisiana, the Carolina Banks and coastal marshes, and every other coastal state, nation, and island were devoured around the world. But such changes and more are already entrained.

Now, again, glaciers globally are melting after a long period of relative stability and sea level is rising at an accelerating rate. There has been an eight-inch rise in the last century. A rise of between one and four feet, by recent estimates, or if the contribution of West Antarctic melting is included, ten feet or more, is expected by the end of the century.[20] As we deny the seriousness of the climatic disaster and make only token concessions to reducing greenhouse gases, we commit the world to melting more and more of the glacial ice into the next century, not so far off. Civilization has expanded over virtually all the earth. People have favored coastal zones, where great cities have been built, connected by sea travel and dependent on a diversity of interlinked marine and terrestrial resources. The continental drainage basins feed rivers, whose floods and floodplains have for centuries supported agriculture, and now support hundreds of millions of people, all increasingly vulnerable to storm-driven floods and rising seas.

The Greenland ice cap, now melting more rapidly than scientists had anticipated, contains as much as twenty feet of global increase in sea level. A recent satellite survey of the volume of the ice showed a current rate of melting equivalent to one millimeter of sea level in two years or about twenty inches in a century due to Greenland alone.[21] The West Antarctic ice cap, also melting, contains another twenty feet. (The Antarctic Continent contains by far the greatest volume of ice.) In a world of seven to ten billion people, with populations densely settled close to the continental margins, a sea level increase of five to ten feet in decades will be awkward to the point of murderous. Anything remotely approaching one to two hundred feet, half to most of the glacial ice existing now, would substantially put this civilization below sea level.

Again, can we defend a policy of accommodation to climatic disruption as it occurs? Just how? Who? When? Where? Those who call for such adaptation see advantage in doing nothing. They should share the circumstances of those who live on the exposed shores of Florida, Louisiana,

Mississippi, the Carolinas, New Jersey, or New York. Or try the Ganges Delta, open to the vagaries of the Bay of Bengal. Millions live there, dependent on small plots vulnerable to any change at all, and impoverished in a crowded world, with no place to go when the sea rolls in after a storm in the Bay of Bengal or the sea level rises just ten millimeters. Adaptation? They can die.

Does the rest of the world owe them consideration? It is that simple. More or less comfortable accommodation to rising seas in one part of the world is death for thousands in another.

Then there are the island chains of the world. Consider the Maldives of the Indian Ocean, the Andaman Islands of the Bay of Bengal, the Marshalls or thousands of other islands of the South Pacific, the Bahamas and the West Indies, and the Philippines, already devastated. They are all occupied. They are all cherished as home by some who have an interest in the earth no less dear than the wealthy corporate masters of the world who prate adaptation as a comfortable and less expensive solution than attempting to reverse a commercially attractive business. Do we not owe island and other coastal dwellers consideration? Were they born without a birthright to life and a place to live and resources to live by? This issue is not a commercial judgment; it is a moral imperative that we demand of ourselves as owed to our fellow citizens.

* * *

A personal family history on Ocracoke Island brings the issue closer to home for me. Ocracoke is a sand strand just south of Cape Hatteras on the North Carolina coast. The village of Ocracoke, with a population less than a thousand, is on the sound side of the island, protected from the open sea by the island, a mile wide at that point. The groundwater in the village, a mere six inches or so below the surface, rises and drops with the tide. The village is flooded regularly by storms that drive enough water from the sea through Ocracoke and other inlets into Pamlico Sound to flood the village feet deep. It was flooded by Hurricane Earl in 2011, and by Sandy as it passed by offshore in October 2012. The next island south is Portsmouth, once a thriving coastal fishing village, but abandoned now for over thirty years after more than two centuries of occupancy. Too wet. Six inches of sea level rise will be the end of Ocracoke Village, too. It's coming.[22]

Not far away on the North Carolina coast, just north of Ocracoke, lies Hatteras Island, the largest of the islands of the North Carolina Outer Banks.

Hurricane Irene in fall 2011 cut two new inlets across the island, isolating the village of Hatteras, and requiring a major effort to fill the inlets with sand and restore the island and the road from the north to the village. This was not the first time Hatteras has been cut in recent years. Such storms, now more common, come with a cost of billions of dollars taken from the public coffers to restore such eroded, low-lying shores—and those are but part of the costs.

* * *

Heat alone can be a plague. Living systems dance to climate, the tempo influenced by temperature, light, and water. Small changes in temperature can have profound effects on who and what does what and when. The difference between frozen and unfrozen is a fraction of a degree Celsius, and that difference can be life or death. The rules that apply to the process of biotic impoverishment discussed in chapter 2 apply here: small-bodied, rapidly reproducing organisms respond most rapidly to more favorable conditions; large attached and immovable organisms suffer, not simply from climate but also from a shifting micro-biological world of competitors and disease. Rising temperatures affect humans, directly through physical effects and disease, and indirectly through effects on the plants and animals, and the plant and animal communities we depend on. The web of cause and effect is complex and intricate in detail, but in the context of systematic impoverishment, predictable. Whole forests succumb to regional warming as species of trees become stressed and bark borers move in. So go the forests of southern Alaska. And the whitebark pine, *Pinus albicaulis*, the famous tree-line stands of the Rockies, a food source for bears and other creatures, is already stressed and disappearing over large segments of its ancient range.[23]

In France: The summer 2003 was hot. Unusually high temperatures prevailed throughout Europe, but especially in France. The high summer temperatures put all life at risk, and human mortality soared, especially among the aged and the poor. A careful comparison of mortality rates over time revealed about fifteen thousand excess deaths in France between August 4 and 18 that year. The total excess mortality in Europe from this unusual heat was estimated as more than thirty thousand, making it one of the ten-deadliest natural disasters in Europe in the last hundred years and the worst in the last fifty years.[24]

But the records of 2003 were mild by comparison with those for the summer of 2010. On July 26, Jeff Masters summarized the experience in just a portion of the Northern Hemisphere to that date in a widely quoted blog:

A heat wave of unprecedented intensity has brought the world's largest country its hottest temperature in history. On July 11, the ongoing Russian heat wave sent the mercury to 44.0°C (111.2°F) in Yashkul, Kalmykia Republic, in the European portion of Russia near the Kazakhstan border. The previous hottest temperature in Russia (not including the former Soviet republics) was the 43.8°C (110.8°F) reading measured ... on August 6, 1940. The remarkable heat in Russia this year has not been limited just to the European portion of the country—the Asian portion of Russia also recorded its hottest temperature in history this year, a 42.3°C (108.1°F) reading at Belogorsk, near the Amur River border with China. The previous record for the Asian portion of Russia was 41.7°C (107.1°F) at nearby Aksha on July 21, 2004....

Six nations in Asia and Africa set new all-time hottest temperature marks in June. Two nations, Myanmar and Pakistan, set all-time hottest temperature marks in May, including Asia's hottest temperature ever, the astonishing 53.5°C (128.3°F) mark set on May 26 in Pakistan. Last week's record in Russia makes nine countries this year that have recorded their hottest temperature in history, making 2010 the year with the most national extreme heat records.[25]

Arguments emerge, even among scientists, as to the immediate causes of environmental change and the significance of the threats. Are the continental droughts at this moment due to human-caused warming or part of a larger pattern outside human influence? Scientists, while recognizing broad patterns, like to keep issues open for further interpretation as new insights and experience emerge. In this instance, however, the broad pattern of continental warming and increased aridity is clear, and the firsthand experience over more than fifty years provides compelling evidence that policy based on adaptation and dreams of "muddling through" is wishful thinking in the short run of years, and suicidal fantasy on the scale of a decade or more.

★ ★ ★

Robert Repetto, writing at Yale, has provided a carefully reasoned and detailed analysis of economic considerations in adaptation within the narrow view of domestic US interests. Even in such a rich and versatile nation, adaptation to immediate changes with little attention to addressing larger causes at the local, national, and international levels becomes at first questionable, and ultimately a fool's errand.[26] Corporate and governmental interests join in this strange charade. Insurance for crops and governmental relief programs, for instance, reduce farmers' incentives to take other steps to avoid losses from newly developing climatic hazards. Federal Crop Insurance Program, for instance, saw indemnified losses rise over the eighteen-year period from 1989 to 2007 from $1.2 to 3.8 billion, an average rise of 6 percent per year. Federal disaster relief programs have become more common and larger, Repetto observes, but as currently constituted they encourage farmers to delay adaptation to the changing climate. The changes in federal programs are not biased toward stability but toward delay in mitigation and amplified disruption. Failures in mitigation are already extracting serious costs in agriculture. A recent report from the International Rice Research Institute in the Philippines showed that rising minimum temperatures reduce rice yields by 10 percent per degree.[27] Similar policies in the United States and elsewhere surround flood control. Hazards are appraised retrospectively, not with an eye toward the future. Dikes are too low, and drainage needs are underestimated. The result? Such occurrences as a flood in Pakistan that destroyed crops over thousands of square miles and an unprecedented flood in the Mississippi valley in spring of 2011, even as droughts afflicted the US Southwest and Mexico, central Asia, and the sub-Saharan African nations. These are not new equilibria to which we can adjust and thereby accommodate; they are steps in a progression of accelerating trends rendering the resources supporting the world's still-expanding human population less and less stable.

There is no solution within the context of adapting to continuous erosion of the global biosphere. Every solution ultimately hinges on a rapid restabilization of essential environmental resources followed by efforts to bring human requirements and expectations as well as corporate entities and public institutions into line with the basic laws defining biospheric safety and stability.

The Age of Fossil Fuels (AFF) is coming to an end. The end is certain but the details are not. If we continue on the present course, the collapse will be environmental as a recalcitrant civilization succumbs to the cascade of catastrophes triggered by accelerating chronic climatic disruptions. High temperatures globally will literally burn and flood civilization off the earth in political chaos sometime after the middle of the twenty-first century, even in the time of many alive now.

Alternatively, the AFF might come to an end through a clear, collective, general decision as we discover we can live longer, more comfortably, and more safely without fossil fuels, and make the transition to non-fossil-fueled economies immediately with imagination and optimism and vigor.

A concerted shift should have begun with strong political leadership by 1980, when the consequences of continuing a poisonous addiction to fossil fuels had already become clear. Unfortunately, despite shrill warnings, the nation's political and economic leaders chose to ignore any qualms and deny the reality. The denial has become a blind fixation to many in positions of power and influence that has continued into the second decade of the new millennium.

Such venal dreams are not unique or even new. Strangely enough, the public has indulged a similar fixation on nuclear weapons through all those years. While we knew the weapons were in actual practice virtually unusable, we have continued to brandish them in numbers large enough to turn every city in the world into smoldering, radioactive rubble, seeded with two to three billion rotting human bodies. The mere fact that the Russians have the same capacity has been justification for maintaining our own inventory as though we, and not they, could survive the exchange. And we continue the fantasy today, still competing in capacity for turning civilization into a cinder—a capacity also bolstered by the financial interests of giant and still-growing national military killing machines selling its wondrous non-nuclear capacities to all buyers.

If the nuclear hazard remains a threat, not yet activated, the effects of the climatic disruption are immediate and real. The effects are now chronic, accumulating in intensity and diversity, and they carry forces as destructive of a viable human future as a nuclear Armageddon. Surely confronting the underlying causes of this actual, cascading disaster in the making is worth an intellectual and financial investment equivalent to the threat of an unwinnable, and universally scorned nuclear war.

I celebrate myself, and sing myself,
And what I assume you shall assume,
For every atom belonging to me as good belongs to you.

—Walt Whitman, *Song of Myself*

The terror engendered by nuclear weapons was balanced in part during the two decades following their use in war by the widely shared anticipation of abundant cheap energy from the atom to drive industrial development globally. The focus of attention turned to governmental alliances with corporate industrial interests. It was easy at that time to slip into the view that cheap energy is a universal good, capable of use to cure virtually any political or economic ill.

I, too, was trapped in this popular view, and considered a major subsidy to nuclear businesses, the Price-Anderson Act limiting industrial and governmental liability in a serious accident, as a wise and necessary step into the new world. My experience was limited, of course, although it did at that point include two years at sea on a naval oceanographic vessel, a steel-hulled diesel "ecosystem," entirely dependent on those deep-down bunkers, tanks of diesel oil, energy. Oil made everything possible, not only mobility, but also lights, heat, freshwater, and air in confined quarters as well as compass, radio, ballast, and pumps. It was that energy, thousands of gallons of it, that kept the ship more or less stable and afloat in the North Atlantic storms that we seemed to seek out to test our mettle as we explored from above the monstrous mountains of the Mid-Atlantic Ridge, then new to science and the public. We were not a good example for we were squandering the future, heedlessly extruding our abundant wastes to float off or sink into an apparently limitless sea, one ship among thousands on the oceans with the same callous scorn for simple manners.

Despite the compelling lessons from the weapons tests, the hard news about the biotic hazards of nuclear weapons, and how their wastes accumulate and circulate widely to become hazards to all, the health of the biotic environment was, for the moment at least, set aside by science, government, and business. It was forgotten. The objective was economic growth, and the prospects seemed unlimited.

It proved to be an unfortunate oversight.

The scientific community, at least part of it, participated with enthusiasm. Democratic capitalism seemed to offer nearly unlimited possibilities in construction and engineering, and in bending the world to suit expanding human interests, aspirations, and numbers. Some biologists, however, were increasingly sensitive to the explosion of the human enterprise into a biosphere whose limits were already being tested. After all, thoughtful scholars were realizing then that our civilization spans at most a few thousand of earth's four-billion-year history. Life itself is an anomaly, and humans are a recent arrival, new and potentially short-term guests in a biosphere owned and operated by the joint efforts of tens of millions of species too numerous to count. The evolution of life over the four billion years of the planet's existence, and especially over the most recent five hundred million years, built the environment of all life as we know it. The total habitat of that life is the thin earthly surface: the atmosphere, the oceans, lakes, and streams, the shallow crustal soils, rock, and elsewhere, the ice. That small volume comprises the entire biosphere, the habitat of all the life we are privileged to know. It extends at most a few miles from the oceanic depths to the top of the highest mountains. A few spores may be carried higher in a turbulent atmosphere, and ancient microbial forms lie deep in crustal sedimentary rocks, but for all practical purposes the functional biosphere reaches its limits with Himalayan peaks and the Mariana Trench, the deepest oceanic cut in the world.

Within that realm, life is a wild scramble for survival among perhaps a hundred million separate forms, or species. Ultimately, each survives by dint of the activities of all. And all survive in a milieu of each, sorted, resorted, and formed into mutual dependencies and overlapping interests that keep them together as guilds or simple communities that define the place and environment, and maintain both—a perfect tautology in nature.

On land, the life at a location defines the site. Much of the Northern Hemisphere is forest. In the northern regions it is the boreal coniferous forest, which is circumpolar and the largest forested region in the world. In eastern North America, the eastern deciduous forest is the natural vegetation of nearly a quarter of the continent. In water, biotic guilds, dependencies as simple as krill and whales, exist but the geography dominates, and we name lakes and streams, shallows and deeps, bays and seas and currents on the basis of what is easily identified. On land and in water, these are the communities that have emerged as the operating systems of the biosphere. They are the products of the evolutionary processes that built the human habitat and maintain it now.

Many biologists and others were increasingly concerned in those postwar years that one species of the biosphere's many millions had in a few decades come to dominate, and in fact control without a plan, the future of all that life. The brutal experience with nuclear weapons showed how vulnerable all human works are—cities, nations, and millions of people—to vaporization in the anger of war. But even more significantly, the experience demonstrated that far short of the convulsion of war, the normal circulating systems of the biosphere are vulnerable to corruption by what, taken individually, may appear to be minor releases of our civilization's waste products.

Such dominance by one species would seem to put the structural and functional integrity of the biosphere at the forefront of human interests, the primary concern in building governments globally in support of a growing population and an industrially expanding, and promising civilization. The experience with nuclear weapons might have been expected to offer a clear warning to everyone just how large the human influence had become as well as how tenuous the human hold on a suitable and nurturing environment might be.

Far from it.

As the human presence expanded and economic growth seemed not just essential but also the very mark of progress, it was convenient for governments and burgeoning industries to assume that the environment would take care of itself, clean itself up after use, and require no special attention. The momentum of the political and economic forces had been such that those forces could redefine the public interest and welfare as "economic growth" at any cost, with "the environment" as a kind of luxury that could be dealt with at a later stage if necessary. Meanwhile, the incremental corruption of air and water gnawed at the panoply of human rights developed over centuries to the point where only conspicuously fatal municipal disasters such as the smog of London and Pittsburgh of the 1940s and 1950s could move governments to action in regulation.[1]

The response of many in the scientific and conservation communities was not at all surprising. More comfortable with small matters that they could control, they turned away from government and business to talk among themselves, accumulated friends who would listen, and built separate nongovernmental institutions designed both to operate independently in conservation and to pressure governments along the way. They had no difficulty in attracting interest. Human interest and imagination have ever been easily captured by the diversity of forms and functions that throng the earth.

For many conservationists, attention in the second half of the twentieth century turned to protecting vulnerable species. Experience with losses of uniquely spectacular forms had seared the human record and a systematic approach to correcting the trend seemed an obvious necessity. The extermination of the fearless and docile Steller's sea cow (*Hydrodamalis gigas*), the largest sirenian and a true giant, reported to reach as much as thirty feet in length and several tons, stands as an early scandalous tragedy, impossible to ignore. At one time individuals ranged along both shores of the North Pacific. By 1768, just twenty-seven years after the first description, the species had been hunted to extinction. Scores of other examples of extinction followed, including the infamous loss of the North American passenger pigeon (*Ectopistes migratorius*), hunted in large numbers for food until there were no more, with the last one dying in a Cincinnati zoo in 1914. In the decades following two World Wars in the twentieth century the expansion of the human presence globally put many other species, including whales, fish, birds, large mammals, and various plants, and still isolated human communities in conspicuous danger.

The response of thoughtful biologists was direct: a listing of endangered species, fauna, and flora, known as the "Red List." It was a compendium of the world's most threatened species published at regular intervals by the International Union for Conservation of Nature (IUCN), an organization established in 1948 in the heady postwar years of hope and innovation. It was founded to expand conservation interests and especially to appeal to the global academic community. The Red List gave formal status to species on an international level that might be used as a basis for public action in protection.

The IUCN quickly became the icon of conservation. Its success and need for funds for urgent new projects led a number of supporters in 1961 to found the World Wildlife Fund (WWF) with an international office in Morges, outside Geneva, Switzerland. The plan was to raise and distribute funds to support IUCN initiatives in science and conservation, and develop the WWF around specific national offices ("appeals"). The US appeal, WWF-US, grew over the course of the next three decades to dominate the world organization. Not surprisingly, the national branches, ever successful in gathering funds from engaged supporters, found it attractive and even important in fund-raising to develop their own programs quite independent of the IUCN. Many innovative programs were created along the way, but the Red List concept easily emerged in the eyes of many conservation agencies as a primary statement of the mission of conservation. A listing on the Red List put the finger on species and specific sites requiring attention and ultimately preservation. And the whole program appealed to the public, which responded generously to requests for funds to save attractive animals in danger.

The approach was effective. By 1963, recognition of the need for more formal support for preserving species against rabid commercial enterprise led the IUCN to draft the Convention on International Trade in Endangered Species (CITES). The convention was designed to prevent commerce in already-known endangered species. By 1973, it had received sufficient governmental recognition that a meeting of representatives of eighty nations could be held in Washington, DC, to establish a process for ratification and implementation of the treaty. CITES entered into force in 1975 after 10 signers had ratified and become parties to the convention. By 2009, 175 nations were parties, and the convention gathered muscle as nations developed their own internal rules for protecting species, such as the United States with the passage of the Endangered Species Act (ESA) of 1973. Additional attention was by then being drawn to special international challenges such as the continued killing of whales despite the International Whaling Commission's decades of efforts to protect them. Big cats and large pelagic fish such as giant tuna and sharks were also increasingly under pressure, and had drawn special attention in conservation.

The emphasis on endangered species was attractive, understandable, practical, and effective in drawing in powerful interest and financial support. But it would often prove inadequate in the core objectives of conservation, even for the species in question. Every species has evolved into a role, large or small, in a community of species. A species threatened is a sign of a community threatened and effective conservation reaches immediately beyond individual species to the recognition of the community and its protection, a much larger challenge that immediately touches land use as well as diverse economic and political interests.

Meanwhile, in the academic world research papers and books appeared around the truism that *extinction is forever*. The Red Lists became longer. The number of species threatened became so overwhelming by the early 1970s that the challenge had been transformed into concern for all species.

A shorthand emerged. "Biodiversity" truncated the term "biotic diversity" and came to consolidate all purposes in preserving species. It leaped over the virtual impossibility of conserving each species, one at a time, and called attention to all life. Defending biodiversity in all its imaginable forms became the core objective in conservation. It was a loose and vague formulation, but economic and political powers find such vagueness often convenient, and the assumption was that governments would surely understand and cooperate. After all, the reference was to all life on earth.

The concept of biodiversity was soon bolstered by books and compendia of papers written by many of the world's best-known scholars asserting articulately the importance to all of preserving every sprig of the earth's green mantle.[2] I was an active participant, eager to advance any effort that would bring greater public recognition to the importance of preserving all the life on earth.

* * *

In 1973, Tom Lovejoy, who had completed a PhD at Yale on bird populations in the Amazon basin, joined the staff of the still-new WWF-US, where I was at that time a board member and chair of programs. There was then great interest in obtaining objective evidence that the metric of biodiversity, the number of species present, defined structure and function in landscapes, and conveyed substance to ecology and succor in various forms to human interests. The search intensified over time as scholars recognized that there are large differences in biodiversity around the globe. Tropical forests, for example, have many more species than boreal forests. "Hot spots" of high biodiversity are obviously more attractive and important as objectives in conservation than places less well endowed. Or are they?

Lovejoy had recognized that the expansion of agricultural activities in the Amazon basin might offer an opportunity to examine the changes in plant and animal species as the area of forest diminishes. In a short time he had arranged an imaginative and aggressive test. He was able to define and establish forest plots of various sizes in an area along the Rio Negro north of Manaus to be preserved while the regions around them were cleared for agriculture. Bird populations in particular were inventoried but records of other species were kept as well over years. It was a magnificent test, and ultimately showed how sensitive such forested lands are to disturbance and what an excellent index of change an inventory of birds offers. All had previously recognized that some forest dwellers such as the jaguar, for instance, range over areas of many square miles, but the sensitivity of bird populations as well as other animals and plants to disturbances had not been widely recognized; further, the demonstrations of losses of species in these experimental plots were unequivocal: preservation of diversity required large areas, thousands of hectares, left intact.[3]

Many other efforts were also undertaken by scientists over the next years to define the significance of mere diversity—the number of species present—in maintaining the structural and functional integrity of communities.[4] Ultimately, such studies would confirm the necessity for preserving large landscapes to assure the continuity of species endemic to a locale. And the species endemic to the locale represent communities whose integrity is essential to other species in the area.

The emphasis on species became a political issue stamped as important by passage in the US Congress in 1973 of the Endangered Species Act (ESA). The act was designed, as the title suggests, to protect species in danger of extinction. Congressional recognition of the importance of protecting species in danger was unquestionably a positive step. It had the unfortunate connotation that only species recognized as interesting and endangered required attention. Overlooked were the issues of land and water management critical to stabilization of essential resources, including drainage basins and coastal land and water. To politicians and the public at large it may have also reinforced the notion that conservation could be set aside as a secondary issue in political and economic affairs, to be considered only on special occasions when a species was at hazard. Then, conservation interests, at least in the United States, had to be placated. And for conservationists, there seemed no other legally powerful way to raise issues of biotic conservation before government. That was an unfortunate and inconvenient fact—a serious stumbling block to effective governance in protecting essential landscapes and water bodies. Its inconvenience and ineffectiveness led to an effort favoring far more comprehensive legislation aimed at improved environmental planning and management in the United States, at the time a world leader on such issues.

One form of this effort emerged as a special report prepared by the National Research Council of the National Academy of Sciences in 1993, twenty years after the ESA. It was a bold and appropriate, if belated, step in coupling preservation of the integrity of natural communities as opposed to species to management of common property rights to clean air, water, and a wholesome, supportive, and vital biosphere. The proposal was for the United States to establish a "National Biological Survey" similar to the US Geological Survey in the Department of the Interior. Its mission would be to identify intrinsic biotic resources and recommend management procedures for their protection. It was a much-needed innovation in government in the United States, a potential model for the world, at a time when the world was increasingly enshrouded by giant international corporate machines and the human population was rapidly closing in on a total of seven billion with the prospect of billions more.[5]

The US Congress had a different perspective, however. So many members of Congress were, one way or another, influenced by exploitive business interests that there was no serious consideration of the proposal. Further, the very success of the ESA had generated sufficient friction in the economic growth community of business politics that there was no chance that any further potential biological restrictions, even those clearly aimed at preserving common property interests such as a forested drainage basin feeding a water supply, would survive. If there were vital issues of conservation involved, they had to be seen through the lens of direct hazards to people or species under the Endangered Species Act, which many political and economic interests also opposed but had to accept. That interesting and progressive piece of legislation, cramped in context and potential as it is, has nevertheless survived, and emerged as what may be the most critical tool in the United States for managing forests and land over the years since its enactment.[6] Its great strength lies in the reality that preserving a species requires that the habitat be preserved intact. Its weakness is just that fact: a single species draws ire if it is the full reason for preventing a lucrative real estate development. The issue writes the text of the contradictions intrinsic in the destructive expansion of the human enterprise in a finite and vulnerable biosphere whose integrity of function is essential to the continuity of the human enterprise.

* * *

"Biodiversity" has, nonetheless, become part of the vernacular, used often in discussions of conservation to refer to all life in a place. Places of high biodiversity, rich in species, are of greater interest for conservation, of course. But establishing a park to protect high biodiversity in a special place is futile if the critical aspects of the regional environment erode and the park slips into progressive impoverishment. The protection of ecological integrity or "conservation" reaches far beyond a simple emphasis on protecting species or biodiversity in parks and reserves.

The regional and global environmental transitions now under way are moving the environment out from under each individual—not species as such, but each organism, each survivor of the competitive business of life, and every special genetic strain, or "ecotype," which virtually every individual represents. In such a world, saving a hot spot of biodiversity is little more than a hope, certain to be ultimately dashed if chronic changes in environment accumulate and produce the inevitable impoverishment of land and water. That set of challenges, so lucidly set before the world by the decade of nuclear weapons tests and the subsequent lessons from pesticides, places the protection of global chemistry firmly in the realm of both conservation and human welfare. Both are the business of government— perhaps its most important business.

Elizabeth Kolbert in a recent book vividly illustrates the importance of global chemistry to species preservation in describing her personal adventures in following scientists to experimental sites on Australia's Great Barrier Reef and to a carbon-dioxide-rich fumarole off the coast of Italy. The research she depicts confirms powerfully the causes of the progressive impoverishment of coral reefs as the seas are warmed and made increasingly acid by the continued infusion of carbon dioxide from the atmospheric accumulation.[7] The reefs are unquestionably hot spots of biodiversity, now rapidly collapsing globally.

Biodiversity became the criterion for establishing an interest in conserving places, whether on land or in water. Hot spots of biodiversity were identified, and parks and reserves were established to preserve such sites. The emphasis on local parks seemed appropriate, as did attempts to connect parks with corridors left in their natural state for the benefit of migratory and otherwise wide-ranging animals. But again, the emphasis was typically narrow and specific. The parks, absolutely necessary and appropriate, all too frequently appeared to mollify the conservation constituency while leaving the rest of the world open to business as usual. Business as usual was in fact destroying all life on earth while scientists and conservationists quietly lived in their own intellectual ghetto, "successful," winning some battles but losing the war. I recall well a conversation with Russell Train, long after he had taken on the presidency of the World Wildlife Fund, in which he agreed that he had never thought we conservationists were doing more than slowing the destruction of nature. At that time and ever since I have thought that objective inadequate. I am sure he was in fundamental agreement and shared the frustration.

To be successful conservation had to become a prime purpose of government, a core effort. The objective had to shift from simply setting aside land or water in parks and reserves, or forbidding through the ESA this or that new development, to restoring and preserving the essential biotic character, the core natural communities of each region. The hazard was and remains erosion through chronic disturbance. But too many of the practitioners were preoccupied, trapped by their own success in advertising the Red List and the losses of eye-catching species. They were caught in their own attractively compelling arguments that "extinction is forever," and in developing the prestige and reach of their own not-for-profit organizations, which at best remained on the periphery of political power. Doors to the halls of political power were open to corporate exploiters, who knew their business and turned to government to favor growth and profits. Conservation, if it were to be successful, had to become competitive with corporate interests, not only in establishing parks and reserves, but also especially in showing that human welfare and the human future depend on preserving the continua of species in natural communities that are the operating systems of the biosphere.

Those insights are not simple. While the technical arguments advanced in the interests of preserving biodiversity were appropriate, and establishing parks and reserves remains honorable and essential, the focus on species was and remains inadequate to the point of being misleading. Any species exists as a population of plants or animals with an array of mutual dependencies in one or more natural communities. The population has a geographic range, large or small. Within that range, the genetic composition of the species varies. Each individual that survives has a genetic constitution appropriate for survival in that place and, quite possibly, not in another place even within the total range of the larger population of the species. The individual is a unique genetic strain, an ecotype. A hot spot of biodiversity may be fascinating for its diversity of species, but "saving" it is saving artifacts unless major efforts are under way to see that there are significant other related communities preserved, and the physical, chemical, and biotic integrity of the region is preserved as well.

Meanwhile, the juggernaut of economic and industrial development rolls over more and more of the residual natural communities and pushes the preservation of biodiversity to the margins. The urgency of preserving natural communities as the maintenance system of the human environment has been lost in the rhetoric of biodiversity, left in the hands of the specialists. Some have found a useful shorthand in framing "conservation" as the maintenance of biotic diversity, yet the concept offered no implement, no tool in science and no model, no example of success, and no compellingly clear cost of failure. It was a concept, attractive but ephemeral to most of the public, and not easily defined, measured, or conserved in practical, specific, understandable terms. Substantial efforts such as the proposal for a National Biological Survey were made to correct those deficiencies, but they proved feeble in the face of economic interests.

* * *

In many cases the establishment of reserves and even rules have been far too late, even to provide a moment's respite from the rasp of industrial expansion and exploitation. The examples are legion and part of all our lives. I recall, for instance, as a boy visiting family friends who lived on T-Wharf on the Boston waterfront. The throng of commercial fishing boats that crowded that large dock several vessels deep enthralled me. These were small, sturdy, diesel-powered wooden craft rigged with trawls that were put over the side and drawn aboard amidships: "beam trawlers" they were called, and I was fascinated as any boy would be. They were replaced in later years by larger "stern trawlers," which continue to this day to work deeper waters offshore. Back then, on family visits to the coast of Maine, I could watch the trawlers working the inshore fishery, and hear the comments of the locals about how the trawlers were taking everything and there was now nothing left. Perhaps it was hyperbole that there was nothing left, but the cod have now been gone from the inshore waters for a long time. Protected zones where no fishing of the species is allowed have recently been established, but it may be too late.

Years later I came back to New England to live in the marine community of Woods Hole, a village on a southern peninsula of Cape Cod where Spencer Baird had settled to investigate fish and fisheries in the 1870s. Convinced that the fisheries were in peril, Baird persuaded Congress to establish the US Fisheries Commission and became the organization's first commissioner, serving without pay. He was witnessing the late nineteenth-century intensification of the fisheries, big traps along the shore, the shift from wind to steam power, and the advent of trawlers that worked the inshore stocks of herring, mackerel, and cod along the New England coast.

Because cod grew to large size and could be "flaked," dried, salted, and traded far and wide, they were especially valuable and their abundance was legendary.[8] Early settlers on Cape Cod described them as so plentiful that cod could be scooped from the water in a bucket. The inshore strains of those cod stocks, despite their original abundance, were fished out and have never returned. Such populations, experience suggests, were ecotypes, genetically attuned to the inshore environments, temperatures, depths, currents, and food, and able to survive the assaults of all predators—except for the trawls. One might think the populations from the deeper waters would replace the stocks. Perhaps, in time, but those populations, too, are ecotypes, fixed to their special places and roles in the world.[9] The offshore populations are under pressure from the diesel-powered stern trawlers, which are larger, more efficient in their catch, far more powerful than their predecessors, and able to stay out for days or more if necessary, hauling larger trawls. Conservation? Biodiversity? Great thoughts, outside the discussion, forced out by the cries of fishermen as stocks declined and fishermen found themselves not only competing for diminished stocks but also at odds with efforts to reduce catches to preserve a residual community of cod capable of recovery from the harvest.

So, too, the once-huge cod stocks of the Grand Banks of Newfoundland were destroyed by overfishing and have not recovered despite many years of protection—and hope. Again, it was clearly the responsibility of government to protect not just the species but also the entire community of plants and animals of the banks that supported, among other species, a huge population of cod. The public interest in a long-term harvest was lost as the Grand Banks of Newfoundland—once, with Georges Bank a hundred miles off the New England shore, the richest fishery in the world—were driven by relentless commercial exploitation into a dismal impoverishment, persistent despite decades of attention and warnings.

The major residual fishery along the coast of eastern New England and eastern Canada in the first years of the twenty-first century has been, not surprisingly, low in the food web. The scavenger populations of lobsters have thrived in the much-reduced structure of the marine ecosystem as the populations of their predators, including the cod, have been greatly reduced. It is a sad, sad story, missed by those ignorant of what was, and now firmly attached to lobsters on the coast of Maine and New Brunswick as a culinary attraction.

The solution was and remains a fierce, effective program for preserving the biotic structure of the oceanic waters. Preservation at this late stage of exploitation requires restoration, assuming it is possible. In either case, the ruthless harvests have to be avoided over large areas set aside to provide a matrix of structural biotic integrity. These reserves must reach into every region: the oceanic surface communities, the deep waters, coastal, oceanic, bays and reefs, and tropical and polar. And they must include explicit protection of the chemistry of all waters. It is a large order, but there is no other way if the ocean's biotic resources are to be preserved. A human population of seven billion and still growing cannot survive long without a closely measured, effective effort to rebuild and preserve the global marine and coastal ecosystems.

The structure and exploitation of terrestrial ecosystems is quite different, but the prevalence of specific ecotypes reigns there as well. Forests prevail in influence on the biosphere by their area and massive stature, carbon content and metabolism. The forests, too, exist in a world of subspecific genetic specialization in which each forest stand represents the local survivors of the relentless selection that is the universal circumstance of living systems. As populations are lost to harvests and impoverishment from intensified disturbances, so ecotypes among all the species of the forest are also affected, not soon to be restored.

* * *

The search for a better rationale for conservation than species preservation has brought suggestions that biodiversity is essential for "the public service functions of nature," the normal metabolic processes of natural ecosystems that keep the biosphere operating. The concept is sound. The integrity of biospheric functions requires all species, and we can describe those functions as having financial value, as they do. Soil and actively flowing streams commonly restore sullied water at no explicit public cost and great public advantage. Where water must be filtered mechanically, the filtration plants can be expensive. Such natural "services" have status and value in the world of economics and politics, and the purpose in assigning a value is to claim that status in the competition with other financial interests. There have been proposals that such services be treated as commodities and sold as "ecosystem service units" available to those who find themselves in need of credits to make up for having destroyed such services elsewhere.[10] The cap-and-trade programs take advantage of the value of certain efforts in conservation or waste management to trade excesses with needy parties, often with a financial consideration. The Regional Greenhouse Gas Initiative described in chapter 6 for reducing the total greenhouse gas emissions within the northeastern states of the United States makes a market in credits for renewable energy among power companies forced by governmental regulations to reduce the use of fossil fuels. The electric power generated by the solar panels on my roof is being sold as credit to fossil-fuel-burning corporations that have not yet changed their power sources sufficiently to comply with new regulations.

While there are circumstances where there is clear advantage in assigning values to natural public services and using those values to protect them against commercial destruction, the monetary approach fails when the public values rise toward infinity, as often happens. In that case governmental control is required.

We live in and exploit for all aspects of our sustenance an environment that is a biotic system. It is dependent for its continuity of function on the totality of its elements, its total genetic pool, not only those recognized as species, but also all of its ecotypes, guilds, and communities of plants and animals, on land and in the global waters. It is the totality of that life that is now threatened with systematic impoverishment—a process that is vividly before us as the biotic feedbacks of the climatic disruption take the potential for control of climate out of our hands.

The enemy is chronic disturbance—cumulative, relentless, irreversible, physical, chemical, and biotic disturbance that moves the world systematically down the scale from cod and haddock to the lobsters and shellfish, down from the forest to the fire-seared blueberry barrens of eastern Maine. Impoverishment proceeds, place by place, until the effects fuse, and the disturbance becomes global and feeds on itself.

The cure starts with recognition that the physical, chemical, and biotic integrity of the biosphere is at issue, and preserving it must be seen as a call to arms because civilization hangs in the balance.

III

WHICH WORLD?

In a society governed passively by free markets and free elections, organized
greed always defeats disorganized democracy.

—Matt Taibbi, *Griftopia*

The bomb, Fukushima, the Gulf of Mexico, Haiti, a jump from two billion
to seven billion people, and a global climatic disruption.

Disasters, all of them—great human disasters of this time on earth.

They span seventy years and more, but only a moment in the million-
year development of the human presence. It is the moment when the
human enterprise ran up against the limits of the global environment, the
biosphere, setting in motion a storm of destruction of essential common
property resources, capped by a rapidly advancing climatic disruption that
promises environmental chaos and a centuries-long accelerating rise in sea
level that might soon be a foot or more per decade with two hundred feet
in reserve.

The response, as we have seen, has been mixed. In the case of the bomb,
new ventures have been limited and efforts continue to reduce the hazards,
yet the hazards persist and grow as new nations acquire weapons. Fukushima, and BP in the Gulf of Mexico were products of corporate industrial
overreach, carelessness, and lack of regulatory control by governments.
They are only signal disasters, large and well known. They are accompanied
by countless lesser accidents and "sacrificial zones" committed to industrial
purposes and profits, and lost to public purpose, for the moment at least.
Haiti stands as example of a failed state, so thoroughly impoverished as to
require massive outside aid to feed its millions. No program exists to restore
self-sufficiency to the nation. The nation is another necrotic spot in the
biosphere, one of many, crying to be cured.[1] At the global scale, expanding

human numbers only intensify the urgency of restoring the biophysical integrity of the earth.

What can governments do? What *should* they do? No answer is simple. Precedents are few.

The Western world has adopted various forms of democratic capitalism as the most practical and successful model of government. They have taken this route in preference to, or even to escape, religious leadership or various forms of central control under dictators or committees, whether popularly elected on not. Among democratic nations, the United States stands as a secular model of a democracy with a strong focus on civil rights and a governmental system that can be made to work by political leaders of competence and goodwill.

The US perspective was powerfully yet simply set forth in the oft-quoted introductory phrases of the Declaration of Independence, penned a mere 250 years ago, but drawing on a perspective with a long history embracing religious and despotic experience:

> We hold these truths to be self-evident, that all men are created equal, that they are endowed by their Creator with certain unalienable Rights, that among these are Life, Liberty and the pursuit of Happiness.—That to secure these rights, Governments are instituted among Men, deriving their just powers from the consent of the governed.

Implicit in this statement is the concept of a human birthright, the expectation at birth of a right to life, a right to clean air and water, food and a place to live in health, peace, and equity with other citizens throughout a lifetime. Such expectations are not unique to the United States. They are shared widely, and celebrated by parliamentary democracies and even, at least in name, many nonparliamentary governments as well. And they are the hope and expectation of billions.

"That to secure these rights, Governments are instituted.... " In the contemporary world of international business, the declaration's words still sound appropriate, if simple and archaic. But there are new actors in this drama. Inanimate new bodies called "corporations" that have acquired rights of their own by dint of judgments of the US court system—rights frequently competing with or displacing those of people.[2] Yet the model set by those introductory words is universally admired and celebrated around the world. The model has, unfortunately, been compromised.

The issue came home to me sharply years ago on a visit to the town of Mexico, Maine, on a morning after a nocturnal temperature inversion had captured the gaseous wastes of the local paper plant and smothered the town in an insufferable odor. I learned quickly that the odor was to many of the inhabitants "the smell of a job" and far from insufferable, although no one believed that breathing such stench improved one's health in any way. Cleaning up the emissions would cost money and the jobs might move south. The state's interest in protecting the quality of air locally was tempered by the need to keep industrial jobs at home. Effects on human health, by contrast, seemed distant and uncertain, and thus not compelling. Nevertheless, all in the town suffered from the noxious air, not just those who had jobs in the paper plant. The corporation had become more than any citizen and could, in fact, usurp the rights of all others without effective objection.

The challenge of cleaning up a factory in Mexico, Maine, was, as in similar cases everywhere, very much an issue of rights and governance. Governments, at least all legitimate governments beyond the ABOTIP (A Bunch of Thugs in Power) stage, take various routes in defining and defending what constitutes the welfare of their citizens.

Rights, rules establishing fairness in dealing with one another and in managing common property for the advantage of all, usually emerge first in such analyses. They have been set out explicitly in the Universal Declaration of Human Rights, adopted by the UN General Assembly on December 10, 1948. The Declaration recognizes and defends "the inherent dignity and ... the equal and inalienable rights of all members of the human family as the foundation of freedom, justice and peace in the world." Thirty explicit articles support in detail those earlier words from US history. The statement offers a broad vision as to the core purpose of government in protecting the rights and interests including access to essential resources for each citizen. It seems a reasonable extension of those expectations and responsibilities to say that rules are established to protect each from all and all from each. I am not allowed to poison my neighbor's well nor is he allowed to poison mine. And if there is a business, the business is not allowed to poison mine or anyone else's. Such rights are commonly established, not only in legislation, but also in case law—legal rights systematically defined by court trials. The Mexico, Maine, experience was simply one example of a corporate interest's flexing its economic muscles at the expense of the public, with the tacit agreement of some of the governed and the government. Suddenly, the corporate "individual" has more than equality among other "citizens." Additional rules are clearly needed in such situations to protect universal interests from those who seek more than equality in access to common property.

The need for rules to protect rights to air, water, land, and all other civil rights obviously increases with the increased intensity of all activities in a growing world. And the growth has been real: a quadrupling in human numbers occurred globally in less than seventy years—in one lifetime![3] The pressures were not confined to mere numbers of additional mouths to feed. Technological advances in that period have expanded the capacity of each individual to draw on resources over large areas. Flicking a light switch in New York may draw oil from a Saudi Arabian well to meet that further burden on the generators in New York. Each individual in the industrial world has a larger command of resources and takes more space in a finite world. Expanding aspirations further intensify the pressures as the developing world seeks equitable access to common property resources and living standards long denied. The need for insight and rules, equitably designed, observed, and enforced, becomes acute.

Growth in the human undertaking generates ever more intensive, and numerous interactions between and among individuals and groups. The "groups" may be corporations in pursuit of profits interacting as or with individuals as well as with other corporate interests. More frequent interaction implies more frequent intersections of interests, more intensive competition for resources, and a greater need for rules to govern those interactions. The frequency of such interactions rises rapidly with numbers of participants, surprisingly so. If the interactions are among equally participating individuals, the number of potential interactions rises as the square of the number of individuals following the "N-squared law."[4] That fact alone stands as an inescapable feature of growth, and reveals the silliness of the arguments that growth can or should be coupled with less regulatory oversight and fewer rules. The intensification of business and demands on all resources requires continuous refinement and enforcement of the rules protecting all, corporate interests as well as rights and interests of individuals. That stench in Mexico, Maine, defined for me one of the most critical governmental challenges of the industrial age: how to maintain essential civil rights and build an industrial economy.

It is in the defense of civil rights and protection of common property from corporate or even governmental intrusions where the failures of oversight—and foresight—are now conspicuously threatening. The most egregious intrusions involve the fossil fuel energy business. The phenomenal success of these corporations, now the largest and wealthiest in the world, gives them extraordinary economic and political power. Some of the corporations such as ExxonMobil are larger than many governments. Arrogance on the part of corporations has led to such cynical views as the writer Derrick Jensen's comment that "the primary purpose of government is to protect those who run the economy from the outrage of injured citizens."[5] Corporations often assume and actively assert that fiduciary responsibilities to investors outweigh any other responsibilities, including any inconvenient moral or civic interests. They can and do use their wealth and loyal supporters to enter the political process and bend it to favor their success. The purpose and practice of government is thereby subverted to corporate advantage. Left unregulated, wealthy corporations can buy their way into devouring the global commons and larger interests of the public. Governments extend the corruption by signing trade agreements that strengthen the governmental corporate powers based on the mistaken assumption that corporate profits will redound to the benefit of the nation.

* * *

Here is a list I made recently of environmental topics covered prominently in the news. All were selected because they involve human welfare—economic and political affairs that are profoundly affected by how governments respond to the environmental challenges.

- A litany of failed and troubled states parched by drought and collapsing into chaos, if not war
- Katrina's lasting scars on New Orleans and the Gulf Coast
- BP's oil disaster in the Gulf of Mexico and its effects documented year after year
- Sections of coastal New York and New Jersey consumed, possibly irrevocably, by Sandy, a record storm
- Pakistani agriculture crippled by persistent flooding
- Sections of Kentucky, West Virginia, and Wyoming, barren after coal mining, remain as festering sores affecting nearby land and water
- Alberta increasingly barren and poisonous to birds and others, including people, after mining tar sands for oil

- The boreal forest of Canada and Eurasia afflicted over vast areas by heat, insect plagues, and fire
- Fukushima reactors and coastal Japan struggling after a tidal wave
- The Philippines struggling to recover from the largest and most severe storm in history
- France and Germany, Australia, and parts of North America afflicted by successive summers hot enough to raise human mortality rates significantly
- The central and southern United States tormented by successive tornadoes
- The world's coastal cities and islands suddenly vulnerable to flooding as larger storms rile a rising sea
- Droughts developing and expanding on every continent
- Nuclear power plants, once thought to be a blessing, now seen as another liability as used reactor sites and wastes accumulate, and require virtually eternal, expensive, and hazardous care
- Migrations of millions in search of places to live in health and safety after having been driven from regions affected by drought

The costs of such disasters accumulate and shift governmental and private funds from routine management or developmental projects to emergency repairs. The damage quickly reaches billions of dollars annually. Such events and trends left unchecked assure economic and political chaos as heat and drought tighten the environmental noose in lands already marginal for human survival. None is conspicuously to be cured by corporate inclination or an international agreement stimulating trade; quite the contrary. Immediate remedies require local attention through communal or governmental initiatives. In the longer term, cures require national or international action.

* * *

Governmental purpose merges with corporate objectives. Money flows from corporate coffers to lubricate political or economic policies. It is a strange inversion that is now, at the beginning of the new millennium, changing governments in even the once-democratically governed segments of the world into corporate economic fiefdoms. The inverted totalitarianism described by Wolin (chapter 5) becomes starkly real.[6]

The United States has company. The Harper administration embraced the concept of corporate exploitation, regardless of national or human cost, of natural resources across Canada. Particular interest was in exploitation of the tar sands of Alberta despite the high expense measured in energy and devastating environmental costs of production.[7]

China, too, is crudely engaged in a capitalistic convulsion. Although there have been protests over civil rights and environmental issues, they have had limited influence. Citizens are abundant, cheap, and easily exploited commercially. Craig Simons, an American writer and political observer, long a resident of China, believes China will respond to outside pressures with "compromise," but those nations pressing for reform will have to lead by example on issues of energy and climate. Again, leadership falls to the United States in restoring respect for the biosphere as the basis of all life.[8]

The totalitarian inversion is a fundamental shift in the structure and purpose of government. New assumptions about political, economic, and environmental affairs have emerged from the sheer size and wealth of corporations that are now effectively international agencies that escape many national controls. A powerful driving force of the transition has been the fossil fuel industry with its extraordinary profits. A mineral deposit, largely publicly owned, has been acquired with subsidies from governments to exploit for profit with substantially no limits on the mining, refining, use, or disposal of wastes. Profits have been tightly focused, and costs extruded into the public realm almost without question. Landscapes damaged in mining coal and drilling for oil and gas as well as the waste products from use of the fuels are potentially expensive liabilities that the industry has largely avoided until now. Corporate owners and managers vigorously defend their ballooning profits and protect their interests through influence on politicians and governments. Thoughtful citizens, scientists among them, have long questioned the process, especially the poisoning of the biosphere, but have not been effective in changing the arrangement.

* * *

Fossil fuels and the fossil fuel industry have been seen to occupy a special place by dint of their importance in supplying the energy that drives the industrial world. Their wastes include the gaseous products of combustion, carbon dioxide, methane, and other gases, such as nitrous oxide and oxides of sulfur, all of which have been released at no cost to the industry—an enormous benefit. The extent of that subsidy is only now becoming apparent. It is the world, the biosphere, all of it, in the process of being destroyed by the willingness of the public to allow the fossil fuel industry to dispose of its wastes at no cost into the common atmosphere and thereby destroy the life-giving global climatic system. It is the ultimate subsidy: the whole earth. It is ardently defended by the largest and richest corporations in the world as well as by states fixed on it as their ticket to accelerated economic development. The fossil fuel corporations are uninhibited in asserting their right and intent to continue profiting from it. They are large and powerful enough, and supported by enough venal interests, to drive governments into submission, and subvert civil rights along with the public interest and welfare. Of course, they started early, and people around the world participated with little objection, ignorant or at least heedless of consequences. The consequences are now well known, but the world has great difficulty rescinding that permission.

The corporate industrial encroachment on the commons has an even more directly destructive, conspicuous side that has been tolerated over decades and formally accepted in many circumstances by governments: the production of wasted landscapes. These are corporate sacrificial zones, as described by Chris Hedges and Joe Sacco, such as the mined-out mountains of the US southeastern states, newly mined plains of Wyoming, the Athabaska tar sands of Alberta, and thousands of disposal sites rendered toxic by incidental chemical wastes.[9] They and other similarly unilateral commitments of segments of the biosphere to permanent wastage for industrial convenience or profit, or simply through negligence or carelessness, are often abandoned into public trusteeship. Some, such as the coalfields of Kentucky and southern Appalachian Mountains, are toxic, and spread their poisons into water and air over time, thereby expanding the area of corruption. They, too, represent a public subsidy of business. Threats by governments to require restoration elicit corporate prophesies of prohibitive costs or revenge at the polls.

Meanwhile, the warming of the earth by one- to two-tenths of a degree per decade assures, if sustained over a century, significant progress toward a catastrophic rise in sea level and a hot, hot world, large parts of which will be inhospitable to humans.[10] The disruption of global climate puts the whole earth in the realm of a sacrificial zone, a gratuitous contribution to corporate wealth.

Over the nearly 250 years since the introductory words of the Declaration of Independence were written, through great wars, many trials and tribulations, governmental purpose and method have been defined and substantiated in the US Constitution and laws of the nation. The model of a democratic republic that has emerged with all its wide-ranging strengths and weaknesses can, in the eyes of most citizens, be made to work effectively in the hands of politicians of intelligence and goodwill who are not tied to corporate or other extraneous interests.

The ecologist, untrained in government but terrified by its current failures, sees three realms that require a sharp, new, or intensified analysis—and repair.

First, there is the integrity of the human birthright: clean, untrammeled air, water, and food as well as a place to live in peace, comfort, and safety from birth through the seven to ten decades of life. Such rights of individuals were asserted by the United Nations in the Universal Declaration of Human Rights, approved by the General Assembly in 1948—a vitally important statement early in the United Nations' history.

Second, there should be absolute protection under laws defended by government from corporate invasion of those birthrights or abrogation of civil rights under any circumstance.

Third, there should be governmental controls that set limits on corporate size, and assure corporate purposes that dilute the profit motive with a parallel set of public purposes consonant with rebuilding as well as protecting the integrity of structure and function of the biosphere, locally and globally.

The net effect of that new focus would be to swing the attention of government away from an emphasis on corporate welfare and expansion toward the rights of individuals and a much-intensified interest in building locally self-sustaining communities in a renewable world.

The perspective shifts back to the individual including the individual corporation, which started on this course early as another "citizen" among equals. That "citizen," enlarged, wealthy, and empowered though it may be, retains responsibilities as a citizen with rights and prerogatives consistent with protection of the rights and prerogatives of all.

Limits on corporate size and realms of operation are not only conceivable but also necessary in the public interest. They have been invoked previously by others over years, and require codification and establishment with public service, not stockholder profits, at the forefront of discussion.[11]

The introduction of morality and public purpose to corporate responsibility is not novel. The current world model for responsible democratic governments has incorporated the innovations and flexibility of English law along with our own legal system in defense of the civil rights of individuals. Interpretations of those rights have evolved over time, but in our day we have sought emphasis on equity in relationships between and among individuals, equity in access to essential resources kept safe and pure such as air, water, land, and opportunities to live. These rights also entail protecting the public from depredations by individuals and individuals from depredations by groups, or by the public at large. Rights embrace the golden rule do unto others as you would have them "do unto you" in its various forms, recognized almost universally in religions and by many governments. In law it is known as the principle of *Sic Utere: Sic utere tuo, ut alienum non laedas*: what belongs to you in such a way as not to interfere with the interests of others."

The golden rule has emerged repeatedly in the considerable body of law developed over three centuries around land and water rights.[12] It also lies at the core of all issues of pollution or theft of common property. It applies to individuals as well as businesses and corporations, although business and corporate interests often seek "to strike a balance" between profits and pollution, between closely held parochial interests and the welfare of all. The most awkward topic is energy. A half century of experience, research, observations, and a continuous flow of data have shown that by burning fossil fuels we are interfering with the interests of others, now and for generations to come. But the principle of *sic utere* pertains to many aspects of emerging technologies including agriculture and medicine where genetic manipulations are opening a Pandora's box of potentially highly destructive interventions. In these new intrusions into the genetic order of the world there are serious issues involving the integrity of the structure and metabolism of the biosphere. Genetic introductions into agriculture soon leak out and become a novel part of the genetics of "natural" populations with untold consequences. There are also serious issues of human health in the consumption of crops genetically modified for resistance to herbicides, diseases, and insects or other pests. Toxins, or toxic effects on insects or plants that are inserted into the genes of crops, are not necessarily benign in humans or other organisms. Safety in humans is a serious issue to be determined in each instance, sometimes by years of work. Corporations have large financial interests involved in these modifications, and routinely dismiss casually, or simply ignore, issues of environmental and human safety that might affect profits. One of the most potentially troublesome intrusions is the combination of genetic innovations that provide resistance to herbicides in crops. Crops carrying that resistance can be raised without mechanical tillage, a giant saving in energy and labor recognized as no-till agriculture.

I recall a conversation in Louisville, Kentucky, not long ago with two farmers, each of whom was farming a thousand acres or more with no-till methods. They were advocates, unequivocal enthusiasts. It was difficult not to envy their confidence and comfort with their roles, much as one fell into a dream when talking with the young rancher from the Amazon basin who flies his own plane among distant farms, each farm a giant by the standards of most. No-till agriculture, a major innovation, is profitable and understandably popular. Unfortunately, the heavy use of herbicides not only contaminates water, air, and crops but also causes the evolution of herbicide resistance in the target species. It is difficult to see how such chemical farming can survive the thickening cloud of intrinsic contradictions.

Growth in all facets of human activities is putting extraordinary demands on governments. Politicians, not necessarily envisioning a public interest beyond corporate profits, are pulled in various directions in an economy they admire as capitalistic, wealthy, and expanding. While most will agree that the business of government is the public welfare, definitions of what constitutes welfare wander widely. Forgotten in this melee is the core fact that the enduring, leading, and essential purpose of business is financial gain, not public service. Also overlooked, and quite parallel to that corporate purpose, is the fact that as growth tightens all demands on a finite earth, the requirements in protecting the human birthright, universally recognized and defined as purpose in government, become increasingly demanding.

Still, in the eyes of many, the health and welfare of the public is not necessarily distinguishable from the health and welfare of corporations. Elected officials, ever requiring popularity for reelection, not surprisingly take an interest in the welfare of their corporate benefactors. Boeing, a giant firm based in Seattle that builds aircraft and sells them around the world, threatened to move a large piece of its operations to Texas. In response, the state and the city of Seattle recently offered Boeing a tax abatement of $8.7 billion and other inducements to stay.[13] Such adjustments are popular because large companies are large employers, but the arrangement constitutes a public subsidy paid for by higher taxes from all other taxpayers, many of whom might also benefit from lower taxes. Improving Boeing's profits with public funds seems questionable at best. One might wonder whether an $8 billion investment might carry with it "fiduciary responsibility" to see that the profits of the business are shared equitably with the investors. Or it might be considered a loan. In the case of Seattle and Boeing, Texas was a competitor, presumably also with attractive blandishments at public expense. But propriety must rule in Texas and other states, too, where citizens may still have standing.

* * *

The needs for governments to establish rules that make a business environment stable, functional, and honest, and taxes to support both government and infrastructure, are intrinsic to society, and grow rapidly as society expands in size and complexity. To argue the reverse is simply to deny biophysical facts and social reality. The rules that are the realm and purpose of government should not only be consonant with the laws of nature (science), but also have a basis in the arithmetic of growth as well as moral expectations applied to the corporation itself.

An emphasis on the individual in the US Constitution and its amendments has been present since the beginning. The concept of the individual, however, has been expanded before governments to apply to corporate groups. The word "corporation" was invented to give the group individuality as a legal entity. That claim has been expanded over time to give corporations not only recognition and protection accorded individuals as citizens but also to claim further influence and rights by dint of working in the public realm and controlling flows of money. Corporations enter the public pasture with many cattle, and by dint of size and money, control the pasture and influence governmental rules to favor their own narrow interests. Virtually every industry open to governmental regulation in any aspect of its business has sought favorable rulings. Examples abound. I mentioned the struggle over five decades to regulate smoking. That struggle continues with the companies busily reaching out to capture youth as new smokers in other nations, where regulations are less stringent than in the United States.[14] The DDT issue dragged on for nearly thirty years. And of course there is the current dance on the energy issue, now delayed half a century to preserve corporate profits.

The Reagan falsehood that "government is the problem" has been the mantra of political conservatives in the United States for three decades. Its adherents argue that regulations are awkward, expensive, unnecessary, and cost profits and jobs. Major efforts were made to reduce governmental regulation of business. A foolish deregulation of financial agencies, including large banks with international business, resulted in the financial debacle of 2007–2008, when "derivatives" constructed by pooling questionable real estate loans proved valueless. Billions of dollars in public money were supplied to avoid the financial collapse now recognized as the worst depression since 1929 from becoming even worse.

Governmental regulations of course need not simply restrain business behavior nor be arranged to favor particular lobbyists; they can be used to enhance common interests by building strength into existing businesses or enabling new undertakings. There are many examples. Land use in drainage basins used as water supplies is commonly regulated to protect the quality of the water. So, too, governments can encourage improvements in efficiency of energy use in household appliances such as refrigerators, thereby making a new market in highly efficient refrigerators.[15]

As the inversion of governmental structure and purpose has proceeded, financial considerations have become preeminent, and used to justify devoting large segments of the earth exclusively and finally to narrow commercial purpose such as producing fossil fuel energy for profit. The explosion of fracking for gas and oil has spread drilling wastes more widely than ever, and contaminated groundwater supplies over large areas. The surface mining of coal around the world has left thousands of acres devastated on every continent. And in producing or using the energy, it seems to matter not at all if the welfare of citizens is impugned. Neither corporate interests nor the politicians they support seem to hesitate to promote policies they deem in their self-interest that corrupt the public environment and result in human morbidity and deaths. They become agents of "malfare," in the language of economist Herman Daly.

The BP disaster in the Gulf of Mexico triggered an immediate explosion of outrage that the safety of workers, the integrity of the marine ecosystem, and the public welfare could be so casually scorned by a large corporation.[16] Unfortunately, the 2011 report of the National Commission on the

BP Deepwater Horizon Oil Spill and Offshore Drilling offered no recognition of the necessity for a retreat from such ventures not only to avoid such indelible intrusions but also to start the restoration of the biosphere.

Even better would have been a stunningly candid governmental announcement, prompted by the oil spill, that the industrial world has reached a turning point and must put a stop to the further development of fossil fuels. Not only can we not afford to risk such an environmental disaster in the Gulf, there is neither "space" in the atmosphere for the wastes nor is there "place" in the world to commit to further increments of biotic impoverishment. The 1992 Framework Convention on Climate Change, ratified by the United States and virtually every other nation, provides every basis required for unilateral national action. But instead, after an appropriate delay in honor of those killed, operations resumed in the Gulf, while elsewhere commerce never stopped but continued with governmental support to expand the production of coal and oil and gas globally.

Industrial intrusions into the public sphere of course do not stop with energy. From the standpoint of human mortality the worst acute industrial affront to date has been the Union Carbide disaster in December 1984 in Bhopal, India, when isocyanate gas and other toxins leaking suddenly at night from a large underground tank killed in excess of three thousand immediately with many more, possibly in the thousands, dying of related injuries over time.[17]

Agricultural and industrial chemical poisoning of the world is far more common than we like to know. Globally, industrial tragedies involving sickness and death occur by the thousands week by week. And then there are the ongoing business-as-usual chemical corruptions. Coal-fired power plants emit through their stacks not only the common waste gases dominated by carbon dioxide but also oxides of sulfur, nitrogen, black carbon soot, and mercury from the burning coal. Distributing these substances across a landscape continuously for decades is obviously not a public health benefit.

Since the 1970s in the United States, federal legislation protecting air and water has been remarkably effective in limiting chronic emissions from industries. Mexico, Maine, among many other places, is no longer plagued by an insufferable stench after a nocturnal temperature inversion. But we are far from the safety and security of people and even farther from protecting plant and animal communities from destruction by chronic exposure to exotic chemicals, to say nothing of the looming threat that increasing carbon dioxide emissions pose in the atmosphere.

The reluctance of government and industry to back away from fossil fuels might be more understandable if there were no viable alternative. In fact there is an array of attractive alternatives that are safer, simpler, and less expensive than mining, refining, and transporting fossil fuels in any form, and they leave no residue to poison climates globally.

* * *

The US embrace of capitalism has produced a clash of objectives with political and economic aspirations often obscuring environmental reality. Extranational corporate interests have pushed for "free" trade to take advantage of cheaper labor and unregulated economic and environmental circumstances in other nations. "Cheaper" labor may result in lower prices, but it also results in competitive labor practices little short of slavery in those nations. The lack of environmental regulations merely transports filthy practices and corporate pollution to other nations ill prepared to combat the wealth and political powers of giant corporations that appear to be offering jobs that would not otherwise be available.

The clash is the more acute for the confounding of economic objectives and environmental security. Poisoning the water of other nations is as corrupt as poisoning the water at home. Bhopal-type disasters are gross derogations of responsibility no matter the circumstance. Issues of civil rights, human health and welfare, and environmental integrity and safety are no less critical in India, Burma (Myanmar), China, and other nations; they only lag behind Europe, the United States, and some other industrialized countries in perception and sense of urgency. The disparities between nations are not corrected by commerce, only exploited and perpetuated. They can be corrected by public recognition of the actual human costs and public pressure for communal action through governments. Meanwhile, commerce rules; the rich rise, and poverty is perpetuated.

In the United States it was not always thus, and need not be in the future. A quite-different perspective, for example, prevailed, perhaps surprisingly, even during the first years of the Nixon presidency, from 1969 into the early 1970s. Palpable levels of urban smog, polluted rivers, and contaminated land had helped spur a substantial grassroots environmental movement demanding legislative correction. The US Congress became sufficiently aware of its duties in defining and protecting fundamental human rights to take serious action. With massive support from the scientific and conservation communities, Congress set a model for the world in the form of several exemplary laws, including the National Environmental Policy Act (NEPA), the Clean Air Act and the Water Pollution Control Act. The Council on Environmental Quality was put in the Executive Office as a part of NEPA and legislation confirmed the Environmental Protection Agency. These were landmark legislative decisions, much needed and widely admired. The steps were enthusiastically endorsed by the Carter administration (1977–1981) as well as by the array of nonprofit institutions focused on human rights and environmental welfare that also emerged at that time and played a major role in stirring congressional action.

Challenging regressive assumptions about environmental issues such as those of the Reagan and Bush administrations that followed has been intensified for the last four decades as committed scientists and conservationists have entered the political fray in the United States and other Western nations. In the United States, these interests have been encouraged since 1913 by a brilliant innovation in the tax structure that facilitated the development of tax-exempt "nonprofit" institutions tapping an almost-universal wish among the public to work in the public interest. Schools and universities fit this class. But so do independent, freestanding institutions that make it their business to define and defend civil and environmental rights and laws such as *sic utere*. These agencies have learned to use the court system to force governments not only to obey their own laws but also to write protective new ones appropriate to the evolution of technology and public needs.

Corporate economic interests have not accepted such innovations in government passively. They, too, have developed institutions devoted to advancing their own parochial interests in protecting business from regulatory controls that might limit profits. The American Petroleum Institute unabashedly defends the further development of oil as a continuing source of energy despite the evidence of destructive effects on climate, health, and other essential human interests. The American Legislative Exchange is a corporate effort to neutralize objections to corporate domination of government, heavily financed by wealthy individuals and corporations.[18]

Contemporary industrial agriculture, as we've seen, promotes the use of chemical herbicides and genetically modified strains of plants resistant to both the herbicides and insect pests, as noted above. The herbicides are sufficiently stable to appear now as common contaminants of water around the world.[19] There are no data confirming the safety of human consumption of the herbicides, exotic chemical structures in the plants bred to be resistant to it, or to the genes used to confer resistance to insect pests. Getting such data would require massive epidemiological studies similar to the early studies of the effects of smoking by Hammond and Horn mentioned in chapter 5. And the reaction of the industrial interests would be similar: systematic rejection and denial based on assertions of scientific inadequacy and uncertainty ostensibly requiring much more research. It matters little whether the issue is genetic modification for agriculture or commercial releases of mercury, lead, or other poisons that have valuable industrial uses; the basic US assumption is permissive until there is evidence of harm. Then, regulation is focused on protection of the human food chain, not environmental chemistry in general. That approach is ultimately self-defeating: a general contamination of natural chemical cycles ultimately reaches all life, ours included. Again, government must become active in protecting the largest public interest, guided by the legal principle of *sic utere* along with the general principles of biology, biophysics, and law.

Concern about the momentum that had developed around economic development at any cost in the postwar years brought forth fierce interests in conservation and human welfare over the final decades of the twentieth century and the first decade and more of the twenty-first. The conservation law movement in the United States over the past forty years produced a revolution in the government of the republic. Powerful new conservation agencies brought science to bear on governmental regulatory activities, even to the point of participating in writing new laws and defending old ones. Organizations such as the Environmental Defense Fund and the Natural Resources Defense Council NRDC emerged with those missions, while scientific institutions such as the World Resources Institute, Stockholm Environment Institute, Worldwatch Institute, and others have joined in exploring environmental rights in an ever more intense and demanding corporate industrial world.

Nowhere are those needs greater than in dealing with civil rights in a world threatened with the torments of virtually unlimited carbon dioxide and other industrial wastes and agricultural chemicals. Regressive corporate interests will be smothered in their own progress if we can manage environmental affairs appropriately—the topic of the next two chapters. The world need not crumble before corporate demands to maintain the status quo to preserve the current model of a corporately dominated industrial world that is progressively engulfing and destroying the biosphere and all in it.

Quite the contrary. The world can turn to recognize and defend the human rights celebrated in the words of the initial documents leading to the founding of a "new nation, conceived in liberty and dedicated to the proposition that all men are created equal."[20] And the World can celebrate the United Nations' expansion of those thoughts in its Universal Declaration of Human Rights, and hold corporate interests to the same standards of citizenship to which it holds all other "citizens."

One touch of nature makes the whole world kin.

—William Shakespeare, *Troilus and Cressida*

Before electronic aids to navigation, back when a navigator stood on the bridge and took sights on fixed landmarks to determine the ship's position, as the landmarks disappeared over the retreating horizon and the last "fix" had been established, it was said that the ship had "taken departure," leaving behind the chaos of recent adventures on land, and setting forth for a new port guided only by compass, the stars, and confidence in the course and destination. The world is clearly in a troubled state: uncertain of our bearings, drifting onto shoals certain to bring fearsome political and economic difficulties. It is time for a new reckoning and new departure with a new destination clearly defined and eagerly anticipated.

The objective is clear enough: a biosphere that will continue to serve as a wholesome and resilient human habitat for the indefinite future. The new departure requires a course toward the closed-system self-sufficiency of an earlier time, before the human capacity emerged for global transformations. The big issues are global: chemistry and climate. While there is no aspect of either of these crises that would not be ameliorated by limiting the human population, the current environmental trends can and must be corrected. Success will advance the conspicuous necessity of limiting human numbers.

* * *

In an imaginative quest for insights into global human affairs in the continuing saga of the occupation of Earth, Gus Speth and Peter Haas imagined themselves arriving for the first time from a distant planet and asked just how they might proceed with settlement. Their earlier planet had been impoverished by the expansion of population to 6.5 billion people, to a point where health, welfare, and possibilities for the future had been eroded, and the time had come to look elsewhere. Earth seemed promising. They decided that the first step was to learn how Earth works as a biophysical system, "a huge science project [to master] the science of environmental sustainability." The central point was that the rules for proceeding in the development of civilization on the new planet had to emerge from clear knowledge of the biophysics of the planet. That in itself was an innovation: the principle that political and economic objectives should be based on scientific insights.[1]

They also recognized the necessity for starting with a set of elemental rules as to how to behave in pursuit of those developments. Speth and Haas chose the 2002 New Delhi Declaration of Principles of International Law Relating to Sustainable Development, but they might just as well have chosen the much simpler core principle of *sic utere* derived from the biblical golden rule for legal use in law.[2] *Sic utere tuo ut alienum non laedas* is usually interpreted as "use what belongs to you in such a way as not to interfere with interests (or use of property) of others." The principle underlies the rationale—if too little the practice—of governance and business at all levels, personal to international. In all instances the central point is mutual protection of self and others—a central element in the human birthright.

The scientific insights into the nature of our biophysical world reach back in recent time to principles of ecology as set forth by Darwin and Marsh in the middle and later years of the nineteenth century. Those insights have been amplified and twisted in various directions over the past century and a half of informed scrutiny but have not changed in their fundamentals. They have in fact become widely recognized, amplified, and adopted as basic tenets of conservation and environmental management. The core insight of the century of Darwin and Marsh was that the biosphere is the product of biotic evolution, tried and tested over aeons, and as outlined in chapter 1, populated by the survivors, species, that together exist in communities that define regions, climates, and biogeochemical circumstances globally. The story of how we have used those insights is not pretty, despite all that we have learned.

The corruption of the chemistry underlying the communities of life and the biosphere as a whole is a core issue in the current crisis. The biological mechanisms for concentrating, or discriminating against, chemical elements and compounds also work with, or against, exotic chemicals (as we saw with DDT) that leak from, or are spread by, industrial activities. A focus on hazards to humans and protection of the human food web is appropriate, and in some cases, effective in spurring remediation. But in the larger context of the chemistry of life, emphasis on direct hazards to people overlooks the wider importance of pathways through the natural world that assure the progressive corruption of air, water, and land—and the ultimate corruption of the human food web. Chronic, cumulative, physical, biological, and chemical changes in the environment quickly undermine whole ecosystems. Anoxic, lifeless zones for example plague the Gulf of Mexico, fed by a Mississippi river rich in nitrogen draining from the vast agricultural lands and municipalities of the central North American continent. These necrotic areas in the offshore Gulf of Mexico are a result of the decay of algae whose abundance is caused by the excess nitrogen. The area afflicted is large—several thousand square miles according to recent studies.[3] Similar problems afflict estuaries and bays throughout the densely populated world.

Each of these dead zones arose from local actions that have destroyed the integrity of local landscapes, lakes, and streams. Industrial agriculture is a major source of contamination of air and water, not only in the Gulf of Mexico, but also globally. The contaminants include the substances used widely in controlling pests or aiding crop growth. The common assumption is that the chemicals used have a short residence time in nature and decay into innocuousness. Some do. But the magnitude of their use and distribution are alone enough to be of concern. Further, the chronic presence of chemical pesticides causes the evolution of new strains of the "pests" that are resistant to the treatment.[4] The problem is large. The total production of all "pesticides" in 2007, the latest year for which data were available, was estimated as 5.2 billion pounds including 2.1 billion pounds of herbicides.[5]

The entire contemporary context of industrial agriculture in using pesticides is corruptive of the normal cycles of the biosphere. The problems associated with DDT use are neither denied nor celebrated but conveniently and persistently ignored when it comes to other pesticides, while the author of *Silent Spring* is, with few exceptions, universally admired. The pesticides industry continues to produce largely unregulated exotic chemicals for "control' of pests because the innovations are easily commercialized,

highly profitable, addictively simple to use, and difficult to regulate effectively and often perpetuate the need for pest control. While biological pest
controls would be far safer and are certainly preferable, they are not easily
turned into a commercial product, and have little support in the commercial market. The extra-industrial effects of the chemicals frequently used in
industrial agriculture are another expensive subsidy paid by the public in
the form of the systematic disruption, and even corruption, of air, water,
land, opportunity, and welfare, often far from the point of use and far from
those benefiting from their use.

Not surprisingly, global cycles of air and water are contaminated. Polar
bears, for instance, top carnivores of the Arctic, are not only vulnerable to
mercury, as mentioned previously, but are also highly contaminated with
agricultural and other industrial toxins including polychlorinated hydrocarbons and brominated flame retardants—all from lower latitudes thousands of miles away.[6] Other mammals, fish, and people share the burdens
of contamination. People are also top carnivores, and those who live in
high latitudes share the vaporous and water- and fat- soluble substances
released elsewhere that persist and are condensed and accumulated in food
webs of the cooler regions. Here we have yet another expensive, although
unmeasured and largely unrecognized, series of subsidies industrial and
agricultural development being paid by unsuspecting citizens far distant
from those benefiting from the convenient, inexpensive indulgence.

Those who dwell in cities are vulnerable to special local hazards such as
the contamination of air. Frederica Perera of Columbia University's Center for Children's Environmental Health has for years explored the influences of maternal exposure to products of incomplete combustion in air,
polycyclic aromatic hydrocarbons. She and her colleagues have shown that
children whose mothers were exposed during pregnancy exhibit "attention
problems and anxiety and depression" that affect academic performance
as well as "peer relationships and other aspects of societal functioning."[7]
Our regulatory system is not protecting human health and is far below the
threshold for protecting natural systems. The issues of toxicity are not transitory problems, diminished by time and experience. To the contrary, they
accumulate and become more serious with neglect. They exist as a result of
ignorance mixed with studied indifference in the quest for profits and economic "development." At the same time, they reflect fundamental structural problems in our current model of environmental management—a

model based on misleading or false assumptions as to how the biosphere functions in support of life.

But the failure is larger than a simple mechanical mistake. The management of environmental resources has become an economic and political as opposed to a scientific issue. The expansion of the human enterprise has encouraged growth in all aspects of the enterprise, but especially in economic development. While the scientist would insist that the expansion of civilization should preserve the natural order of the self-sustaining biosphere, fixation on economic growth has brought "compromise" or simple exploitation of all aspects of environment for profit. Beneficial economic growth must follow ecological rules.[8] This view of the world is quite outside the current model of limitless global growth based on unlimited global markets. Yet this view is essential if the current global trends in chemical corruption and climatic disruption are to be corrected, and civilization turned from catastrophic erosion toward restoration and vigor.

* * *

While the chemical structure of the biosphere is obscure, and defined largely by experiment and analytic insights, the biotic structure is conspicuous and its function more easily defined. The unit of structure, the first level of analysis, is the community: the normal plant and animal communities of each place.

On land, each community is dominated by plants and approaches the characteristics of a closed system without being in fact "closed." Energy, water, and many nutrients flow through such systems, but there are many internal cycles that assure the continuity of the resources and integrity of neighbors as though living by our principle, *sic utere*, tried and true. All is driven by a flow of solar energy through the plant community, which accumulates essential nutrient elements—carbon, nitrogen, oxygen, potassium, and a host of other elements—and controls their flows.

We can only admire the precision and orderliness of the normal successional march of such plant communities over time—an orderliness made the more conspicuous as exotics drift into the system and occasionally, freed from normal controls in their new environment, sow chaos. One of the most famous of a rapidly increasing number of such exotics is the pernicious southern Asian vine known as kudzu (*Pueraria lobata* and other species) now abundant in the southeastern United States, where it is

a common roadside pest, potentially covering all other vegetation including trees. But other exotics have brought even more pervasive disruption.
A fungus (*Cryphonectria parasitica* [formerly *Endothia parasitica*]) introduced
early in the twentieth century destroyed the American chestnut (*Casta-
nea dentata*), once a major species in the forests of eastern North America
and now extinct throughout virtually its entire original range. (A few trees
remain in a stand in north-central Maine, apparently remote from sources
of infection.) In the forests of eastern North America, bittersweet (*Celastrus
spp*), another pernicious vine originally from Asia, can cover a forest stand
in a few years and kill the trees. There are scores of others, often serious
pests, recognized year by year. A recent introduction, the emerald ash borer
(*Agrilus planipennis*), threatens the elimination of ash trees (*Fraxinus spp*)
from their extensive range in North American forests.

Forest communities can be large in stature and area—large enough to
dominate regional biophysics. The extensive forests of the Amazon basin
discussed in chapter 4 with root systems at depths of sixty feet and treetops
that exceed two hundred feet in height control moisture regimes conti-
nentally and influence climates hemispherically. Their impoverishment
proceeds by a combination of chronic chemical and climatic disruptions,
massive harvests, exotic introductions, and transitions to intensive agriculture along with climatic changes. The world's great forested regions exert
such influence on biospheric function that their impoverishment on a large
scale reverberates through the biosphere to the further detriment of environmental stability.

Extensive research on forests such as that outlined in earlier chapters but
with long-term programs focused on forested drainage basins, especially
those at Coweeta in North Carolina and Hubbard Brook in New Hampshire, have offered the most useful insights into the structure of terrestrial
ecosystems and their capacities for controlling the qualities of water under
various management regimes.[9] So it is understandable that we turn to forests for insights into the effects of climatic disruption. The impoverishment of forests in the climatic debacle not only deprives the world of an
array of continuing biophysical resources in support of human interests
but also releases significant further carbon into the atmosphere from both
the decay of trees and the heat-accelerated decay of organic matter in soils.
Conversely, their management to favor human interests in environmental
stability and safety is essential.[10]

Regional energy budgets reflecting flows in and out, and water, and nutrient budgets on land are commonly viewed as simple physical or mechanical resources, inherently stable, but all are heavily influenced by life processes, especially forests. The forested regions of the normally naturally forested zones function as a great biotic flywheel whose momentum stabilizes regional biophysics to the benefit of all life. Forests have enough influence that their restoration and preservation over the normally naturally forested portions of earth is a major objective in restabilizing the human habitat.

The oceans present a different story, powerful in that the energy exchanges between the surface waters that cover two-thirds of the earth and the atmosphere affect global climates continuously. But in contrast with forests, the biology of the oceans has a far more modest influence on the global carbon budget. The most important influence at this moment in the evolution of the climatic catastrophe involves the approximately two billion tons of carbon that is annually the net flux from the atmosphere into the oceans. That carbon enters the complex oceanic biochemical system and increases acidity to the point where organisms that make carbonate shells are affected, as mentioned previously. Some of the carbon dioxide that diffuses from the atmosphere into the surface water of the oceans is taken up immediately by algae, and started down the chain of transitions leading to sedimentation and removal from circulation, at least for the present. That transition over vast oceanic areas is limited by various other factors such as nutrient elements, especially nitrogen and iron, required to support large populations of algae. These changes in the oceans are serious and threatening in that they involve major shifts in the populations of those plants and animals low in the structure of marine food webs that support all the larger marine organisms, fish and mammals.

* * *

Along with reversing the trend toward increasing corruption of nature's chemistry by closing industrial systems, as suggested in chapter 2, the reversal of current global climatic trends is clearly critical to any chance of returning to a resilient and lasting human habitat for the indefinite future. "Reversal" means just that: the establishment of a cooling trend to return the earth to approximately the global climatic regime of 1900, when the carbon dioxide content of the atmosphere was three hundred ppm by volume or slightly less—a concentration that had not been exceeded in hundreds of thousands of years until recently, when human activities started the progression to the current four hundred ppm, rising one to three ppm per year. The establishment of such a reversed trend will raise many questions as to how far to go and how to control it. But would not those questions be far more acceptable than the current threats of a runaway heating of the earth?

The prospect of such an abrupt change in direction has not arisen in sage political councils of governments and has not yet been considered in the nominally scientific councils of the Conferences of the Parties (COPs) to the Framework Convention on Climate Change (FCCC). The FCCC, after all, in 1992 set the much more modest goal of "stabilizing" the heat-trapping gases in the atmosphere. The twentieth COP to the convention in Lima in December 2014 was the first to open intensified activities toward the goal of stabilization since the ill-fated efforts of the 1997 Kyoto Protocol. Those intensified efforts had been encouraged by direct action by the United States, first, domestically through the EPA by President Obama's forcing a major shift of electric power production away from coal and toward renewables over the next years—a major step in reducing the largest source of US emissions. Second, Obama made an agreement with China, now the world's greatest emitter of carbon dioxide, to join with the European Union in new limits on emissions. China agreed to cap its emissions by 2030 and to attempt to reach a peak earlier while also increasing its non-fossil-fuel energy to about 20 percent by 2030.[11] While these agreements represented giant steps by comparison with previous efforts, they fall far short of what is required to avoid climatic catastrophe as outlined above. Agreements were also made to continue the discussions, thereby raising hopes that the COPs will finally be effective.

Further delay among the world's major greenhouse gas emitters will bring additional climatic troubles that eat personal wealth as well as national wealth and well-being. There is no escape from the necessity of a rapid movement away from fossil fuels globally. Reaching the point of reversing the current trends and cooling the earth will require early, substantially immediate abandonment of fossil fuels coupled with persistent efforts through a century or more to lower the heat-trapping gas content of the atmosphere. It can be done. Controlling carbon dioxide is the primary objective, but methane, nitrous oxide, and other heat-trapping gases cannot be ignored.

US leadership on all these topics is both essential and effective, as the agreements have shown. While the shift to a non-fossil-fuel economy cannot be made overnight, there is within the biotic causes and cures of the climatic crisis surprising potential to ease the transition.

* * *

Forests are so large in carbon content, area, and metabolism, locally and globally, that there is no chance of correcting current trends in atmospheric carbon without managing forests. The key observation compelling this conclusion is the annual oscillation in the concentration of carbon dioxide in the atmosphere (see chapter 4). Over a period of six months in concert with the seasons, the carbon content of the atmosphere is changed hemispherically by as much as 2 percent or more. The concentration declines during the summer months to a fall minimum and rises through the winter to a spring maximum. At any given time, the carbon content is an integration of the effects of the photosynthesis of all green plants and the total respiration of the plant and animal communities of land and water in each hemisphere. It is a huge transition involving ninety to a hundred billion tons of carbon globally. The amplitude of the oscillation is higher in the Northern Hemisphere, where there is more land, and in the higher latitudes, where the seasonal differences in light and temperature are most pronounced. The oscillation calls attention to the significance of the vegetation and its metabolism in affecting the atmospheric carbon content even over a period of weeks.

Forests are especially important because they cover a large segment of the earth, have a large green-leaf surface area (three to five times the land area), and store large quantities of carbon in trees and soils. Forests at one time covered about 44 percent of the land area and currently less by most estimates.[12] Deforestation by harvest or fire, or to clear land for agriculture, releases the stored carbon into the atmosphere, largely as carbon dioxide. Conversely, the restoration of natural forests restores both the land and, over time, large pools of carbon in the vegetation. The potential of forest metabolism for removing carbon from the atmosphere and storing it in trees and soils is large, several billion tons annually.[13] Conversely, the destruction of forests is potentially a huge additional burden of heat-trapping gases.

The total annual global carbon emissions as a result of human activities in these early years of the new millennium are approximately 10 billion (10^9) tons of carbon (about 30 billion tons of carbon dioxide) and rising. This amount is the global sum from burning fossil fuels, about 8.5 billion tons of carbon, and deforestation, about 1.5 billion tons. Approximately half that total, 4 to 5 billion tons, accumulates in the atmosphere annually.[14] The remainder is absorbed into the oceans, about 2 billion tons, and by plants on land, 2 to 3 billion tons.

The extent to which carbon dioxide is removed by these two processes, absorption into the oceans and uptake by plants globally, appears to depend on the atmospheric concentration of carbon dioxide. That conclusion emerges from the observation that the removal of carbon dioxide from the atmosphere annually has remained close to 50 percent of the emissions over six decades even though the total annual emissions have risen substantially. An increased absorption by the oceans is understandable as the concentration in the atmosphere at the air/sea interface rises. The increase in uptake on land is less easily explained. It requires an increase in carbon fixation by plants generally. That could reflect an annual increase in the area of forest, which seems unlikely, or an increase in photosynthesis unbalanced by a stimulation of respiration. It seems reasonable to assume that the absorptive capacities of oceans and land would continue at approximately the same value, at least initially, if the total emissions were reduced. Can we take advantage of that momentum and hasten a reduction in the atmospheric burden?

The immediate objective of climate stabilization has already been established and agreed to by the nations globally under the 1992 FCCC: stabilize the heat-trapping gas burden at levels that will protect human interests and nature. Nothing more specific was offered. Short-term stabilization requires removal from emissions of an amount of carbon equal to the net annual accumulation in the atmosphere, currently 4 to 5 billion tons of carbon as carbon dioxide. Presumably, if we could remove 4 to 5 billion tons from emissions in one year, the other sinks (the processes removing carbon) would still exist in that first year and there would be no additional accumulation in that year. To maintain that advantage, the emissions would have to be reduced in subsequent years as well. But the greatest challenge would be the first year. What would it take that first year to reduce emissions by 4 to 5 billion tons of carbon?

The 4 to 5 billion tons of carbon accumulating can be offset in toto now by reducing the use of fossil fuels and managing forested land.

Consider the forests first. The atmospheric carbon increment from annual deforestation currently, about 1.5 billion tons of carbon (table 10.1), comes from the destruction of forests, especially in the moist tropics and the north-temperate and boreal forests. The world's remaining primary forests can and should all be preserved as one of the biotic wonders of the earth of inestimable value. Their universal preservation would remove at least 1.0 billion tons of carbon from the currently excess annual emissions, and possibly as much as 1.5 billion tons.

New forests offer even greater potential. In the normally naturally forested zones, a successional (developing) forest of 1 to 2 million square kilometers stores annually about 1 billion tons of carbon. (Alaska has an area of about 1.7 million square kilometers; France and Texas, each, have about 700.000 square kilometers.) So the deficit, the net accumulation in the atmosphere, can be reduced by managing forests, first, by stopping any

TABLE 10.1
Changes in the Annual Global Carbon Budget in Billions of Tons to Accommodate the Agreements under the 1992 FCCC

Total emissions of carbon into the atmosphere annually	~10
Components: Burning fossil fuels	~8.5
Deforestation	~1.5
Residue accumulating in the atmosphere annually	4–5
Potential for correcting by or stopping deforestation	~1.5
Reforestation 1–2 million sq km	1.0–2.0
Residual to be removed from fossil fuel emissions to reach stability now	1.5–2.5

Note: About 50 percent of the total emission (~5 billion tons) is absorbed into oceans and terrestrial vegetation currently. The remainder, 4 to 5 billion tons, accumulates and is the current problem. In the short term that residual can be eliminated with entirely constructive efforts. In the longer term the efforts become more demanding but attractive if constructively approached.

further deforestation of primary forests, and second, by restoring forests on 1 to 2 million square kilometers of originally naturally forested land. Such land is available globally as abandoned, impoverished, or otherwise "unused" areas. Further, forests in drainage basins control the quantity, rates of flow, and quality of water supplies as well as various other regional biophysical attributes essential to human welfare and to the functional integrity of landscapes. In this calculus, management of forests would be worth in total 1 to 3 billion tons of carbon in the 4 to 5 sought, a major step in meeting obligations under the 1992 FCCC. This conclusion, summarized in table 10.1, emerged from deliberations of the World Commission on Forests and subsequent analyses, largely by the Woods Hole Research Center.[15]

Success in realizing those two steps—halting primary forest deforestation and restoring forest—would leave a remainder of 1.5 to 2.5 billion tons of carbon to be removed by reductions in the use of fossil fuels globally in this first step (again, see table 10.1).

* * *

The transition outlined here is essential. The total carbon content of the northern forest, the largest forested area in the world, its soils, and the peats and soils of the tundra exceeds by at least twice the current burden of carbon as carbon dioxide and methane in the atmosphere, or about 800 billion tons. Its total metabolism probably approaches a third of the total metabolism of terrestrial vegetation globally worldwide. The global seasonal metabolism of terrestrial vegetation involves absorption of carbon through photosynthesis and release through respiration of about 90 billion tons. The fraction involving the Arctic and the boreal forests could easily be 30 billion tons. A few degrees of additional warming could cause a 10 percent increase in the total annual respiration relative to photosynthesis and add 3 billion tons of carbon to the atmosphere. The destruction of forests, or of the vegetation and soils of the long-frozen segments of the Arctic tundra, by disease, insects, or fire—all favored by the warming already entrained—also has the potential for large, further releases of carbon. The climatic disruption under way now is moving rapidly into "feedback" systems that can destroy the biosphere. The forests and Arctic may respond in different ways, but the risks of devastating releases are high enough that the additional releases should be avoided at all costs.[16]

Meanwhile, if nations led by the United States can move rapidly, the potential exists in the next few years for slowing or possibly deflecting that tragedy through a combination of managing terrestrial ecosystems, especially forests and their soils, and reducing the use of fossil fuels. The opportunity is likely to be transitory, short lived, and once lost to feedbacks, irrevocable.[17]

On the world stage, among the first moves to reverse the trends in climate should be a quick turn to preserve the residual natural forests globally. They are all critical landscapes: the tropical forests of the Amazon and the Congo, the great forests of the Pacific islands, the circumpolar boreal forests, and the montane forests of the world. All can be set aside as essential in the restoration of the biosphere, now and forever, for there is no substitute for the forests globally in managing a stable biosphere. They are already public treasures to be cherished and admired, a magnificent gift to our successors, a monument to the biosphere that was. But they are also an essential element in maintaining a habitable earth, a segment of the global human habitat whose continuity is necessary.

Such changes in land and forest use and purpose are significant challenges for all governments. Yet most governments do own or control enormous segments of land, just as the United States does through the Bureau of Land Management, the Forest Service, and other agencies. There are also various methods of subsidizing efforts to use private or municipally owned land for essential public purposes. Those subsidies can in some cases be international grants encouraged by the World Bank and other agencies.

Information on land areas potentially available and the current activities on land are increasingly available through greatly improved remote-sensing techniques. Such techniques are now capable of monitoring details of forest cover globally to "keep score" on local, regional, and national progress.[18] The remaining primary forests have become vital global resources, part of the global commons, essential to the welfare of all. There will be endless arguments about profits from the sale of timber and more land in agriculture to meet the demands of an expanding human population. But the age of massive deforestation to feed greed or enable the expansion of industrial agriculture has passed as the climatic disruption generates continental droughts and equally distressing floods in marginal regions. The nations globally came to a quite-remarkable agreement in ratifying almost immediately the 1992 FCCC, and must now implement it by joint and individual actions. The transition will mark a critical turning point if nations can accept that the management of land in the normally naturally forested zones affects not only local interests but also regional and global climates that must be protected in the interests of all.

Important as they are, primary forests are not enough, as noted above. The most effective short-term system for removing carbon from the atmosphere is through the photosynthesis and storage of carbon in plants and soils. Again, the management of terrestrial ecosystems, especially the reestablishment of forests in the normally naturally forested regions, has the largest potential. While the objective is global stability of the atmosphere, the action sought is obviously local action in management of land.

* * *

Throughout most of human history prior to the Industrial Revolution, personal subsistence, even survival year by year, has been tied closely to the resources of place, one's personal locale. That conspicuous dependence has been relaxed only over the last two hundred years as cheap and versatile energy has brought industrial economies that seemed to resolve all human needs and more, at least in part of the world. Cheap energy seemed infinitely promising. Except that it was not. The biosphere itself was pushed into crisis by the large scale of the new industrial world that appeared, superficially, autonomous and overwhelmingly powerful.

Place has emerged again as important. It is the local ecosystems, the essential parts of the biosphere, that keep the whole, the industrial world and the biosphere as a whole, intact and functioning for all life, including people. The crisis of the biosphere can be alleviated but only by attention to the details of place, all around the globe. Forests come first because they are so large in the world—large in area, stature, and capacity for affecting the global carbon budget.

The world needs and can afford a major effort in reestablishing forests throughout the normally naturally forested zones. Forests are not built overnight but successional (developing) forests store carbon. Systematic efforts at identifying regions that can be returned to natural forests simply by abandoning the land to forest succession will establish a new flow of carbon into landscapes. The total annual increment to be gained by abandoning land to forest succession is uncertain. It lies in the several-billion-ton range. Over the course of a century and more at those rates, the forested biosphere can be rebuilt into a broad spectrum of desirable forest systems. If carefully planned, these reforested areas can serve multiple purposes such as extending urban parklands or creating healthy drainage basins for water supplies. None of these considerations avoid the absolute necessity for drastic reductions in the use of fossil fuels, but they do ease the problem.

Restoring the normally naturally forested regions of the earth is a key step not only in in restoring stability to climate but also to the biosphere as a whole in a human-dominated world. How much of the potentially forested regions should be in forest in fact, recognizing the realities of human demands for land and forest products? Any answer is certain to be arbitrary, but the objective is to recognize that the earth is a park and should be protected as such. Forest-supporting climates are limited in the world and should be cherished by restoring and retaining forests as the dominant influence. As a "dominant influence," the naturally forested zone might require land use patterns designed to retain as much as 85 percent of the landscape in forest at some stage of development. The immediate objective is to change the flow of carbon from net emissions to substantial flows into biotic storage. The total area to be set aside would have to be large—as large as Alaska. Such an area, if distributed worldwide in a wave of public interest in global restoration and renewal, might be seen as a great blessing of new parkland. Subsidies provided, for example, by oil- and coal-producing nations, those that have become wealthy from the fossil fuel business, might start the process, although ultimately it involves a general tax burden shared by all. But those costs will be far lower than the costs of failing to deflect the current drift of global climate and systematic biotic and economic impoverishment of the earth.

These two steps—the preservation of all remaining primary forests and an active program of reestablishing forests of former forest zones—successfully implemented within five years or less, would take the world more than halfway toward the immediate initial goal of a year with no further atmospheric accumulation of carbon dioxide. They would also provide major advances in restabilizing landscapes globally, reestablishing reliable water supplies, and supplying cooling shade in new suburban savannas and parks everywhere. It would be a major advance in emphasizing that the biosphere is a park to be preserved as such by local actions because it is our common and only human habitat.

* * *

The immediate objective of removing emissions of carbon dioxide sufficient to have a year with no net further accumulation in the atmosphere
would still require a reduction in emissions from the combustion of fossil
fuels, but the reduction using the calculations above would be 1.5 to 2.5
billion tons, or approximately 25 percent of the approximately 8 to 8.5 currently emitted (see table 10.1). A 25 percent reduction would be substantial,
of course, but entirely possible in the short term through modest steps in
improved efficiency, conservation of energy, and the inevitable transition
to renewable sources. It would require wide acceptance of the necessity
for abolishing fossil fuels rather than, as it appears at present, expansion
of their production and use. Further efforts in avoiding coal, gas, and oil
would be necessary both to maintain the new status quo and still further
reductions to advance the transition. But achieving that first year's success
before 2020 is well within reach, given sufficient political interest and will.

Significant reductions beyond stabilization of the atmospheric burden
will take many years of vigorous efforts at restoring forests, reducing fossil fuel use, and rebuilding soils. The longer-term problem of returning to
about 300 ppm requires removing the carbon that the oceans have absorbed
over the past century and more. While it is a technically straightforward
series of steps to stabilize the atmospheric composition to meet the agreements of the FCCC, further steps toward returning to levels that prevailed
prior to 1900 will be difficult.

The major problem emerges early, immediately after the first flush of
success. As the atmospheric concentration of carbon dioxide drops below
the concentration in the surface waters of the oceans, there will be a reversal in the net flow of carbon to restore the equilibrium. While in recent
years the net flow has been into the oceans, that sink disappears as the diffusion pressure difference is reduced. The carbon transferred to the oceans
over the two centuries of increasing atmospheric concentrations becomes
available again to the atmosphere. It is a large amount of carbon. Currently
the net flow is about 2 billion tons annually into the oceans, as we've seen.
The total over the two centuries in question is in the range of 100 to 200
billion tons or perhaps more. Assuming no further releases from fossil fuels
as we proceed in the restoration of the atmosphere, that total must ultimately be accommodated in addition to the current excess atmospheric

accumulation of an approximately equivalent amount. The oceans, however, are enormous, and while they have been acidified by the added carbon dioxide, the flux back into the atmosphere is slow. The atmosphere mixes hemispherically in days to weeks but oceanic mixing occurs over decades to centuries. The oceans will not recover from the current excursion for a thousand years or more, no matter how effective our storage of carbon becomes. Nevertheless, as discussed, it is essential to make every effort to restore the atmosphere.

Forests and other terrestrial vegetation, including acid bogs and, to a lesser extent, swamps, marshes, and soil organic matter, which include the fibrous roots of grasses and tundra peat, offer virtually the only additional immediately effective way for removing atmospheric carbon and storing it for decades to centuries. Agricultural techniques can favor the accumulation of carbon in soils. Other techniques may be developed in time using renewable energy to store carbon in the earth in various forms, but at this moment, biotic storage is readily available and effective. Success requires a global commitment of land, especially forested land, to the special purpose of biospheric climatic stability and human welfare. The transition is not simple, and requires a revision in the current trends in the industrialization of agriculture as well as in corporate dominance of government.

The most likely source for immediate reductions in release of carbon dioxide of more than 25 percent is the large industrially developed nations where improved efficiency and the shift to renewables would be advantageous. The will to achieve such a reduction must be developed internationally and would need to be as broadly supported and as well focused as the industrial system was in 1939–1941, when the United States and other nations turned from peace to war. The immediate, short-term reductions at issue are simply efforts at conservation that should be taken anyway: eliminating unnecessary lighting, reducing and making more efficient automobile and air travel, and reducing heating at night or other times when not necessary. But the unequivocal change in direction away from fossil fuels is needed now. The time for equivocation has passed.

These transitions will not flow easily, but they offer fascinating, fresh even excitingly attractive possibilities in a low-energy, clean, comfortable, and more nearly stable world—a New Departure to be taken with confidence and enthusiasm.

The Golden Rule tries to integrate the individual to society; democracy, to integrate social organization to the individual.

—Aldo Leopold, *The Land Ethic*

Politics, economics, and environment are the substance of human affairs from the family through the great issues of the global politics of survival in a squabbling world. These are the three core elements of civilization, each dependent on the other two, a trivet supporting all human ventures. Now, in the beginning years of the third millennium, the environment, ever vulnerable to competition from economic interests, weakened and frail, crumbles. The economic interests have gained strength and now carry weight normally taken by governments, whose vigor also wavers. Civilization is at hazard.

The global crisis of environment has, in the West, been generated by the callous expansion of democratic capitalism over the surface of a clearly finite and vulnerable earth. The cure is not more conventional economic growth but a return to fundamentals: the re-establishment of the physical, chemical, and biotic integrity of the ecosystems that collectively make the biosphere. Their conservation and maintenance is at the center of the world's urgent business in re-establishing a basis for subsistence in a multi-billion-person world, already slipping into environmentally triggered chaos.[1]

Lest there be any question, apart from war, wealthy and powerful corporations stand at the core as cause of this dangerous time. Elementary biophysical facts set stringent limits on corporate size, wealth, political influence, structure, function and purpose in an ever more crowded and intensively used world. New rules are needed as it becomes clear that the

environmental leg of civilization's trivet is weak. We might review important observations:

- The **biosphere is the sum of its parts**, each of which is itself a biotic
  system, a local ecosystem, whose physiognomy and biophysical functions must be considered in all aspects of the management of land, sea,
  and air.
- The **n-square law** indicates how rapidly growth in population increases
  the intensity of interactions among people, and assures the need for
  increased rules for protection of common interests in environmental
  security, equity, and human rights (chapter 9).
- Chronic disturbance of virtually any type, physical or chemical, causes
  systematic biotic impoverishment of natural communities, affecting
  their role in maintaining a stable and functionally intact biosphere.
- Extensive experience accumulated around the circulation and hazards
  of radioactive particles and DDT and its residues shows that for such
  long-lived toxins, there is no "safe level" of use, **no threshold** below
  which accumulation cannot occur and safety is assured.
- Corporate and municipal responsibilities must extend to development
  of **closed systems** for both industrial and municipal developments as
  the *sine qua non* of the next decades of adjustment to a world of renewable energy and climatic restoration.

Beyond the poisoning of the biosphere by the waste products of fossil fuels and other industrial developments is the closely allied explosion
of an industrial agricultural system, rife with corporate monopolies, across
the landscape of every continent. It is an insidious threat to the use of
land and water for it is profitable, and it appeals as essential in providing
food for the ballooning billions while it uses its apparent inevitability to
appropriate and dominate the use of land for a single purpose over vast
areas. While industrial agriculture may be commercially attractive, it often
replaces vitally needed primary forests. These large enterprises, made possible by elaborate machinery and supported by ingenious innovations in
crop breeding, are heavily subsidized by governments in the United States
and elsewhere. They have pushed the areas cultivated beyond the margins
of sustainability by mining water at or near the outer limits of arable land.
In Yemen, for example, aquifers used in irrigation are dropping at two
meters per year, and worldwide, according to Lester Brown, "water tables

are falling and wells are going dry in some 20 countries including China, India and the United States—the three countries that together produce half the world's grain."[2] While these systems are frequently capable of introducing inexpensive food in large volume to the market, they are addicted to industrial chemicals including herbicides, insecticides, and nitrogenous fertilizers that leak into water supplies and global cycles; expensive, patented genetically engineered seeds and crops; and expensive specialized machinery. Land ownership and control are accumulated to the exclusion of small farms and the diversity of local specialties in crops and products that small farms cultivate.

This consolidation of land and agricultural production in large corporations is often seen as risky policy in food production as well as socially destructive.[3] Small farmers are not immune to corporate pressures. To have a market, for example, small farmers are often forced to join in new industrial cropping systems that require investments in seeds and equipment on the assumption that the sale of the crop will enable repayment of loans, a large risk in a period of uncertain climate. *Bloomberg News* reported in May 2013 that "more than 2200 farmers in India committed suicide in the past four years as water loss … [and crop failures] drove them … into debt."[4] The large agricultural corporations, however, thrive, having pushed the risks of crop failures, ever more certain as climates disintegrate, onto farmers.

For millennia people have lived on the land on every continent at or slightly above the subsistence level, using local resources. Specialized crops were developed locally and improved over generations by the cultivation and selection of indigenous plants. Small farmers indulged in a diversity of crops using inexpensive cultural innovations that improved yields and assured a variety of possibilities for success. Without design or plan these uses of land were consistent with a clean, enduring, and renewable world. The smaller farms have even recently also offered a place and a living to people who would otherwise have been displaced to urban centers to seek jobs. Now, with the massive expansion of industrial agriculture, the time has come for a new look as accelerating crises of land and water use and pollution with agricultural poisons are joined by a global disruption of climates that threatens to cripple extensive existing agricultural regions.[5]

* * *

An intensified division between the haves and have-nots, the rich and poor, lies at the heart of the global crisis of environment and is devouring the traditional agricultural enterprise. Industrial agriculture is now dominated by wealthy corporate giants far removed from the land and oblivious to the subtleties of a global crisis of environment. Their perceived responsibility is to profits, not to public service, human welfare, or the starving millions. More generally, corporate business, including agriculture, is now, apart from various military ventures, the dominant international economic and political force, cultivated and encouraged by governments as the locus of jobs and the generators of personal well-being.

Where does the public welfare lie in this increasingly global redistribution of land, opportunity, and wealth? In the United States, the separation of rich and poor was intensified recently by the Supreme Court's *Citizens United* decision reversing long-standing and appropriate limits on money in politics. The decision put the energy companies, the wealthiest corporations in the world, far up on the pile, and passed parallel advantages to other wealthy corporations and individuals. Abundant, even ancient, precedents for the financial control and deflection of governmental purpose betray any claim of innocence of the decision's consequences on the part of the US Supreme Court. Plutarch, for example, writing about the Roman Senate in the fourth and third centuries BC, observed

> that the abuse of buying and selling votes crept in and money began to play an important part in determining the elections. Later on, however, this process of corruption spread to the law courts and to the army, and finally, when even the sword became enslaved by the power of gold, the republic was subjected to the rule of the emperors.[6]

Rome survived for several centuries, although hardly as a paragon of civil rights. Now we have the urgent pressures of global environmental collapse virtually hammering on the public consciousness as human numbers soar and civil rights are increasingly threatened. What will work in preserving biophysical integrity and a viable civilization, and thus assure universally the broad spectrum of human rights for succeeding generations? The obstacles are large.

There is, first, no escape from the tragedies imposed by expanding human numbers, which can ultimately defeat any government or plan. The total numbers become a conspicuous issue of personal welfare as limited common property and common interests are divided ever more finely among competitive users. Decisions as to limits on the total population, local, regional, and global, while unpopular in some circles, are an obvious personal and governmental necessity for survival in a world whose human expansion has long since passed conspicuously beyond safe or sustainable limits. Clear acknowledgment by governments of that fact and the rules required to retain civilization with already-dense populations is fundamental to a viable future.

Second, the momentum of dominant economic arrangements is being allowed to increase the spread between the rich and poor, with corporate power and wealth displacing government just as governmental control is most needed. That momentum, recognized, can be checked deliberately by taxation and governmental action as it was a century ago when the income tax was introduced, implemented, and used later as a powerfully leveling influence.[7] The importance of greater equality is clear, brought forth most conspicuously recently in powerful statements such as those by French economist Thomas Piketty and US senator Elizabeth Warren.[8] Despite the obvious need, and intellectual support as well as experience, there is no sign that the current US Congress, largely controlled by reactionary forces, will soon be moved to share such concerns and correct them.

Third, no longer is it reasonable to assume that egregious local destruction will be repaired by the earth's restorative capacities. That simple fact requires a major revision in perspectives on management of human activities and widespread adoption of the concept of locally closed systems (chapter 2). That emphasis, self-sufficiency based on local resources used renewably, is the essential conclusion of economist Herman Daly's analysis of how best to cope with the transition of the world from "empty" to "full."[9] A similar perspective emerges in various forms from others such as William McDonough and Michael Braungart. Not all are optimistic that civilization can respond appropriately, but most see a necessity for what James Howard Kunstler refers to as "decent behavior" with a local focus on management of resources.[10]

Agriculture rivals the military-industrial complex for influence within the United States and other governmental systems. Direct agricultural subsidies in the United States come through the Farm Bill voted on regularly by Congress. The total annually is of the order of $100 billion in subsidies for large industrial agriculture.[11] That bill and congressional elections are both heavily lobbied by agricultural interests, which can and do spend billions in attempting to influence votes. Major US corporations participating in this scramble for wealth and control include agricultural giants such as Monsanto, ADM, and Cargill as well as the American Farm Bureau Federation. These large direct (and indirect) subsidies fix in place corporate purpose in use of land and water in agriculture for much of the most productive agricultural land in the world.

* * *

When the local dairy farmer's man stopped by the other day to cut my ten-acre hayfield, I had an inside look at modern "small" farming in New England. He came with a giant tractor pulling a cutter that clipped a twelve-foot swath about two inches above the ground. The crop was heavy because last winter the same man had visited with a giant tank of liquid manure, which he had sprayed over the full ten acres. In the spring, the grass leaped to life and was four feet tall by mid-June. The mowing may have taken thirty minutes, and man and tractor were gone, stopping only to fold up the cutter to fit the road. The man returned late the following morning with a tedder—large mechanical fingers that fling the hay into the air to speed the drying. Two hours later he was back with a mechanical rake, which cleared a fifteen-foot swath and left the hay in a succession of windrows. Within thirty minutes the rake, too, was folded for the road and the hay left in neat rows over the entire field. An hour later, man and tractor were back, this time with a pick-up baler, a special device that ran down the windrow, packed the hay into rectangular bales of about seventy-five pounds each, and dropped the bales in the field to be picked up later. One man and at least half a million dollars in equipment harvested a bumper crop of hay on ten acres in less than three hours, all from the seat of a tractor and without a single hand tool. It was an astonishing show, and a far cry from the farmwork of my youth and generations before me. This dairy farmer in southern Maine milks about eighty cattle, manages about three hundred acres, and has at least one brother working for him. They are nurserymen, veterinarians, botanists, businessmen, tractor repairmen, and in winter could also even now be lumbermen and foresters managing woodlots for wood and lumber. They have an investment of several million dollars in land, silos, barns, and equipment along with experience and knowledge of a highly specialized family business now in the second generation of that family. These are variations on the time-tested theme of subsistence living, now transformed by machinery and requirements for cleanliness, safety, and regulated competitive markets into a proto-corporate status, but still just a bit above the subsistence level.

Subsistence farming was the primary way of life around the world for thousands of years and for three hundred years after the European settlement of North America. One liked to think an industrious man could start with a hoe and build a farm. At one point, perhaps, but industrialization changed all that in a few decades. Now a seven-billion-person world, plagued with a surfeit of people, finds its environmental base fragile and threatened to the point where political, economic, and agricultural systems start to fail as well. Land in corporate hands offers little recourse to the dispossessed and unemployed, the cast-offs of the Industrial Revolution.

Now my well-established, proto-corporate, specialized dairy farm is changing. It has sprouted large plastic-covered aluminum "hoop" houses and a roadside stand where fresh vegetables, expertly prepared for the market and freshly harvested, are being sold. The farm is becoming rediversified in purpose and product, and meeting a surging interest and a growing market for local produce. Is it the forerunner of the next big change: a rediversification of agriculture and local resurgence in use of the land? Are those solar electric panels covering that enormous barn roof? Is that a big electric tractor being charged in the toolshed? Can it happen at the community level as the advantages in cheap labor elsewhere disappear, jobs again become local, and towns become independent for energy and water, sewage treatment, and much of their food?

Cities, too, will change but much more slowly. Jared Diamond, in his perceptive analysis *Collapse*, has shown how vulnerable nominally well-developed, potentially competent communities are despite clear paths to success and continuity.[12] The Icelandic Danes, for example, who settled in coastal communities in Greenland in the tenth century, were accustomed to the subsistence farms of their youth. Transported to a new climatic regime where survival required a focus on the resources of the coastal sea, they were unable to adopt new models of subsistence and their colonies ultimately disappeared.

Our world has a parallel challenge but has far greater potential for versatility in adjusting to earthly limits. Although emphasis shifts among the political, economic, and environmental legs of the trivet, all three must function responsibly and in mutually supportive ways. War, economic depression, or an environmental disaster such as a severe drought can individually snatch, or even destroy, any hope of progress over decades. But effective intensification, intrinsic in a seven-billion-or-more world, requires the local focus discussed above: a grand improvement on the subsistence farms of the past but in the same bold model. And it requires a plan and a universal sense of stewardship to keep the local balance of political, economic and environmental interests stable and supportive. The prospects, while demanding, are attractive and far from forbidding. As Robert Repetto, summarizing a famous 1984 conference of the World Resources Institute dealing with "The Global Possible," observed, "We have sufficient knowledge, skill, and resources—if we use them. If we remain inactive, whether through pessimism or complacency, we shall only make certain the darkness that many fear."[13]

* * *

There is a large "debt to nature," accumulated over the two centuries of industrial development powered by fossil fuels. That debt must be amortized early and on a global basis. The trend toward increasingly severe climatic disruption was long ago seen as a universal threat and made the topic of the now famous treaty ratified by all nations—the 1992 FCCC. Chapter 10 showed how the world could set forth on a long-term goal of restoring an atmosphere approximating that of 1900. In addition to the role of forests in the management of the climatic crisis, agriculture has a role in storing carbon in soils not widely recognized nor often accepted as appropriate. The inevitable retreat of corporate agriculture from arid zones will force intensification on a more restricted land area. In addition, there will be pressure to avoid cultural techniques that contribute to the current chemical corruption so thoroughly dangerous and widely deplored. The new emphasis on localities, towns, and municipalities as internally integrated, energy independent, segments of the biosphere will play a significant role in encouraging changes in agricultural methods, the diversity of crops, and productivity.

Virtually all agriculture benefits as organic matter, humus, is accumulated in soils. The deep organic soils of the North American prairie, for example, were built over thousands of years by deep-rooted grasses. These are some of the richest and most productive soils in the world. They have been tilled in ways that exploit the carbonaceous soils, releasing more carbon through decay into the atmosphere, and as the organic matter has decayed, soil depths in the natural grasslands have declined. While there is no simple universal way to reverse that trend in carbon, much of the cultivated natural grassland and the semiarid steppe build soil carbon stores through the growth of fibrous-rooted grasses. Roots may extend to depths of many feet. Such highly organic soils are especially well developed in areas of loess (windblown silts of glacial age) such as much of the tall-grass prairie of North America. The carbon retained and new carbon stored through the enhanced growth of grasses can be substantial. Moreover, significant efforts are under way to breed grain crops that are perennial, require little or no tillage, and will build soils year by year as grass crops do now. While there is no panacea here, the topic is already well advanced in research by programs similar to those of the Land Institute in Salina, Kansas, where Wes Jackson and colleagues are developing perennial wheat and other grains, opening additional doors to major transitions in agriculture.[14]

The current trends in accumulation of land for industrial agriculture in anticipation of surging demands for food are widespread and powerful. They are, however, contrary to the focus that must now be made on local self-sufficiency. A joint report by GRAIN and the Brazilian La Via Campesina, small organizations specializing in the transition in agricultural land, observes that small farms still manage 70 percent of the world's farmland. But the pressure is high, according to this report, to yield to "land grabbing corporations who seek only speculation and profits." The report claims that the bulk of food production comes from the high productivity of land held in small parcels and intensively farmed. Nevertheless, corporate interests are expanding their acquisitions. GRAIN notes that "land occupied by just four crops—soybeans, oil palm, rapeseed and sugar cane—has quadrupled over the past 50 years [as] 140 million hectares of fields and forests have been taken over." A World Bank report shows that the lands purchased from smallholders are often not put into production, at least immediately.[15] The paradigm of endless economic growth on huge industrial farms, enabled by the revolution in farm machinery and abundant inexpensive fuels that enable irrigation, is countered, according to the World Bank, by the advantages of local food production on small plots. That consolidation of land and wealth at one end of the spectrum in corporate or proto-corporate hands, a "grow-or-die theory," is the antithesis of what is required now in restabilizing the human habitat to the benefit of all. Restabilization will require a substantial inversion of the worldview of politicians and the public trapped in a "corporate-think" mode.

The "success" of the fossil-fuel-powered industrial agricultural system is, unfortunately, in large degree an attractive myth when examined in the larger context of maintaining a biosphere capable of supporting civilization indefinitely. It follows the corporate model, described pithily by Ambrose Bierce as "an ingenious device for obtaining individual profit without individual responsibility."[16] While the industrial production of food appears to offer to feed the world's impoverished and hungry, the food enters a market in an economic system that does not include the hungry, who are hungry because they are poor and essentially outside the market. And while, with the best of intentions, food may from time to time and place to place be made available through various routes at no cost to those who cannot pay, unless special arrangements are made, supplies are often hijacked and sold. They return to the "market." The best, most reliable, universally available sources of food globally are local producers supplying themselves and local mouths and markets.

Industrial agriculture fails, then, on three counts. First, it leaks toxins that are not and cannot be controlled, threatening natural systems globally as well as human health and welfare. Second, while it is financially profitable, it serves only those who are a part of the industrial economic-political system. Those displaced from the land often become homeless, unemployed, and pushed outside the economic system, hungry. Finally, the expansion of massive industrial agriculture as arid zones, too, expand on a climatically stressed earth puts additional pressure on the forests and forestland essential to the restabilization of the global climate. Clearly, we need a new set of expectations and rules.

* * *

The rules needed are simple enough. They require recognition that it is the natural communities of plants and animals that have defined the conditions of life in all the biosphere, and their essential functions must be protected globally. The transition in outlook is profound; it entails an all-encompassing renewal of interest in and support of the living world. Wes Jackson has long argued that the enduring agricultural model must ultimately be based on and integral to the structure of the natural communities displaced. While Jackson is a conspicuous and vigorous proponent, he is

far from alone.[17] An agriculture that rebuilds soil carbon stocks while producing marketable grains is obviously an urgent objective for this century.
Its pursuit in research should be a major focus of governmental attention.
It is unlikely to emerge from laissez-faire commerce yet it would be a profoundly important economic development.

At the other extreme of laissez-faire ideology is a new view of corporate purpose such as that embraced by Paul Hawken, who suggests that the
"promise of business is to increase the general well-being of humankind
through service, creative invention and ethical philosophy."[18] A similar perspective is embraced by McDonough and Braungart who envision a new
business ethic in closed material cycles as simply good manners as well as
efficiency in the use of energy and materials.[19] These perspectives of course
remain far from the realities today of the larger corporate world's competitive business rules.

Linda Greer, a scientist on the NRDC staff, and her colleagues presented
a startling glimpse into that world. They set out to discover the extent to
which a large industry's noxious wastes could be controlled, potentially
at a profit to the company. They developed a collaboration with the Dow
Chemical Company in Midland, Michigan. Dow produces a wide range of
chemical products including agricultural chemicals. Between 1996 and 1999
Greer and colleagues, along with Dow staff, examined the various manufac-
turing processes within the corporation with an eye toward improving both
efficiency and control of waste products. The major points were recorded in
the following conversation with Greer, published by the NRDC:

Q: Why on Earth would NRDC want to work with Dow, and vice versa?

A: It's like Willie Sutton said when they asked him why he robbed banks: that's
where the money is. If you want to prevent pollution, you've got to go to the
source—that was NRDC's motivation.

  As for Dow—we'd worked on a project with Dow a few years ago where we
identified a number of pollution prevention measures for them, which they then
failed to implement. We made our displeasure over that public, and I think that
was a motivating factor for them in going ahead with this project. I don't think
they actually expected that this project would yield significant pollution savings,
because they honestly thought they'd already done what was there to be done.
But it was a win-win proposition for them. If the project hadn't produced all
that well, they would still have gotten credit for trying, and they would have
continued to argue that pollution prevention wasn't the answer. Since the project
succeeded, they'll get credit for stepping up to the plate—and they deserve it,

because that's just what they did. They had no legal obligation to do this, and it really wasn't just about the public-relations value of this for them. Their local plant managers took this seriously and followed through.

The other very significant players here, of course, were the local activists. They played a role that you'd really have to sit at the table with us to appreciate. These are folks who have been fighting Dow tooth and nail for years over a range of issues, most notably dioxin emissions at the Midland plant. For them, this was personal, because their children drink the water that Dow pollutes. So they saw this as an opportunity both to educate the local plant managers about the daily impact of their operation, and as a chance to help prevent pollution. I think they were successful on both fronts....

Q: How do you feel about helping to save Dow Chemical a bundle?

A: Well, let's just say that if you told me five years ago that I was going to help Dow increase its profitability by $5 million a year, I would have stood and waited for the punch line! But at this point my view is: whatever it takes. I'm certainly not going to stop working to improve the laws. And we'll keep on taking polluters to court, as we always have. But if another way to get companies to stop polluting is to save them money, so be it. With the hundreds of millions of pounds of pollution being released—legally—into the air and water every day, we obviously need to develop supplemental approaches to protect the environment.

Besides, the fact that this project saved Dow money will help us persuade other companies to do the same, and that's the objective.

By the way, just to be clear, NRDC got no money from Dow for this project. We raised money to cover our time and travel expenses, as well as time and expenses for the local groups, through grants from two charitable foundations and from the U.S. EPA and the U.S. Department of Energy. Dow paid for a technical consultant and a project facilitator, and covered all of the Dow staff time that went into the project.

Q: Did the actual pollution reductions and cost savings meet your expectations?

A: They exceeded our expectations. On the pollution prevention side, we set a reduction goal of 35 percent, and, frankly, we thought that was ambitious. But we came in with a 43 percent reduction! To do that in only two years is just amazing, given that fighting for these same kinds of reductions through the regulatory process, which is full of loopholes and delays, can take years. The same was true on the money front. I'd hoped the project would break even. But Dow's already made back what it invested and will save $5 million a year from here on out. It's testimony to the kinds of opportunities that are there to be found, provided someone looks for them.

Q: In light of its profitability, why isn't Dow racing to reproduce this project at its other plants?

A: This was the biggest eye-opener for me in the project. I went into this thinking that if we could find ways to reduce pollution and save money at the same time, it would be a "no-brainer"; we'd be home free. How naive! Examined from the view of a large corporation, the question's a lot more complex.

First, there's the problem that $5 million isn't an awful lot of money from Dow's perspective. They make daily investment decisions ten times bigger than that. So, there's the problem of getting senior management's sustained attention with these small sums. Second, saving that money required a certain investment—just over $3 million. Evidently they have in mind other investments for that $3 million that they think will be more profitable. Finally, for yet-to-be-enlightened business people, this project solves a problem they don't think they have. From their perspective, they have a government permit to treat or release these chemicals from their factory, and they are meeting occupational health standards, so what's the problem? They're busy people, and so you can see why these projects would not be a priority.

Also, while I think plenty of folks at Dow do care about the environment, there's not much in the corporate culture that rewards that sort of concern. Managers don't even get bonuses or promotions for effective environmental work, for example. Until that changes, it will remain an uphill battle.

Q: If other companies follow this model, will environmental regulations eventually be unnecessary?

A: No way. In fact, I think the project suggests just the opposite, because it teaches us just how financially enticing environmental opportunities need to be in order to be naturally appealing to a business. There's much more work to be done here, of course, but in the end, I suspect we'll need additional ways to motivate industry to tap the opportunities that exist for them to radically reduce their reliance on toxic chemicals.

Q: What are the most important things that you learned from this project?

A: Two things stand out. First, because the pollution reductions and cost savings at Midland were so significant, we need to assume that other such ventures would yield significant savings as well. It raises hope for radical reductions in pollution from this nation's factories, if we can ever get pollution prevention to take root. That's very exciting.

Second, and less encouraging, the profitability bar is set higher than I thought it was; motivating companies with "business value" alone is not going to suffice. That is, I'd thought that industry would opt for cleaner production methods if doing so was a break-even proposition. But what I learned is that industry leaders have a lot of inviting things to do with their money and they don't think it makes business sense to invest dollars in pollution prevention when an investment in something else will give greater financial returns. Which means that to be effective, pollution prevention needs to be not just profitable, but more profitable than other potential investments. That's a real challenge.[20]

The corporate industrial world is unquestionably wondrous and versatile. Its powers, however, do not extend to guarding the commons or defending human rights. Just as the editors of the Scientific American saw no responsibility to science, scholarship, or the public beyond keeping the business afloat financially, so the large chemical industry looked continuously for better investments. Cleaning up wastes or marginal improvements in the operating system was not a good investment. Such responsibilities are extra-corporate interests that may be large in the eyes of the public but require governmental regulatory power to appear important in the day by day operations of the industry. While democratic governments use compromise as a universal lubricant, growth in the human endeavor forces ever-intensified governmental action to protect basic human rights. Governmental rules are required to raise the "cost" of noxious emissions to the point where companies such as Dow cannot afford to poison the world and must take an interest in both internal efficiency and meticulous control of wastes. The data in support of those decisions must come from science, not on the basis of ad hoc research, but rather on the basis of firm knowledge of the requirements of a safe and functional biosphere—requirements long proven, well-established, broad rules that are beyond the infinite challenges of commerce seeking to pass expenses into the public realm.

The closed-cycle concept (no leaks) illustrated by and demanded of the nuclear industry, once embraced, is applicable far into the industrial world. The best moment for reasserting and accepting it as fundamental in the structure of civilization, including all of government and business, might well be the time of transition to enduring, renewable clean energy, the end of the poisonous fossil-fueled era when attention is drawn to the worst and most disruptive atmospheric pollutant of all.

If business profits depend on a license to dispose of wastes into the public realm to the detriment of the interest or welfare of others—a clear violation of civil rights—there should be no basis for the business. A corollary is the necessity for corporate responsibility for goods produced and sold into the market. "Responsibility" involves implementing the closed-cycle concept: recovery and reuse or remanufacturing of materials at the end of their useful life in the previous product. In this context, materials and durable goods, whether rented or purchased, are returned to the manufacturer for reuse or salvage.

That vision requires continuous articulation and restatement as the world tightens around a corporate culture that has with impunity overlooked long-established principles of civilization. "Corporations" were invented to convey to collective activities in business a series of specific advantages in the search for profits. Is it too much to expect public responsibilities to accompany those rights? By what logic does a business acquire the right to poison anyone's air, water, or food, and argue before government for the right to do so?

The right, logical or not, is often assumed as a corollary of the democratic system, and may be one of its greatest vulnerabilities. The persistence among members of Congress in denial of the causes and seriousness of the climatic disruption over years has been a sad demonstration of the weakness of our electoral system. Those denials, while couched in pseudo-scientific terms, constitute a derogation of duty in that they have been advanced to prevent action promised and approved by Congress through ratification of the FCCC in 1992. Driving out such behavior will require an electorate capable of putting into office intelligent and responsible agents.

Seeds of the needed transition are sprouting now in the United States. They are small and tenuously viable, but they reach from local food production, water and land use, and forestry, to a cascading shift in energy production, distribution, and use. And they reach into conservation and long-term planning to preserve segments of natural ecosystems as well as protect whole drainage basins as water supplies. Together, they acknowledge the objective of restoring the functional attributes of the ecosystems that have built the biosphere, recognizing in essence that the world is a park, to be treasured for all time. International conservation agencies have long recognized this objective as a necessity but have found themselves petitioners before government, far less powerful than corporate interests. Their issues are now before governments as critical to the continuity of effective political and economic systems. And the winds of change are gathering strength as thousands have taken to the streets of Washington and New York to impress on government the seriousness of action in stopping the climatic disruption. Interesting transitions are under way as well in the return of local farmers to local markets. In Santa Monica, California, I have watched and admired over several years the development of a well-organized and disciplined farmers' market. It started with a few stalls on a single street, and has expanded to a hundred or more stalls on several streets with a thousand customers, all in the center of the city. The goods presented are of superb quality, all local California products, offered by professionals who make a living farming family-owned enterprises, proud of their products and willing to talk. Some farmers come as far as seventy-five to a hundred miles to join the market, open only for hours. Several make a business of cultivating produce that commands a high price, and are equipped to go from one such local market to another and have regular customers in each market. Some take orders. All have mail addresses. It is a new, local, specialized business cultivating local products expertly raised and handled.

In contrast, I joined in a shopping expedition at a giant big-box store on the Peninsula south of San Francisco to buy fresh fruit and vegetables. All were "a product of Mexico," all packaged in plastic and served on forklift pallets in stacks four to five feet high and four feet square, or a half cord by my woodcutter's reckoning. A week later, I surveyed what appeared to be the same raspberries in similar packages, also "a product of Mexico," in a giant supermarket in York, Maine, nearly three thousand miles away. Who could but wonder and admire a system that could produce such quantities of "fresh" berries in markets so remote, and presumably, at many points between and beyond? Yet at what costs? All three markets are, of course, heavily dependent on fossil fuels, cheap labor, irrigation in an arid zone with diminishing water supplies, cheap transportation, and hardy crops that can stand the handling, storage, and ultimate sale. It is an astonishing system that works—for the moment.

But in York, the new farmstand opens new possibilities for commercial self-sufficiency. Some young entrepreneurs came by a year ago seeking an opportunity to develop a large garden on my open land. The giant local grocery that has Mexican fruits has an expanding section for carefully labeled local produce identified in some cases by the name and place of the farm and grower.

The dream that the world, desperately in need of biophysical restabilization, can be celebrated and protected as a park is buoyed by the urgency of the moment and some remarkable previous successes in enlisting communal action for common purposes. The idea lies at the core of beliefs that fuel a thousand or more conservation agencies around the globe including such giants as the WWF, Greenpeace, NRDC, Environmental Defense Fund, IUCN, Conservation International, and Nature Conservancy. And there is more interest emerging daily as the seriousness of the climatic disruption becomes apparent and intransigence of the wealthy corporate interests tied to dirty energy is brandished flagrantly.

There are models of carefully planned controls of land and land use. Norway and Sweden have had land use policies that have preserved their landscapes over centuries. The United States has its national park system and extensive Forest Service and Bureau of Land Management holdings, all nominally under management in the public interest, although "management" has favored local, often commercial interests, even in the national parks. A far better model is the Adirondack Park in northern New York State. The park is a thriving contemporary example of wisely integrated development. It is a 6.1-million-acre reserve, about 50 percent publicly owned, with the remainder in private hands. The largest contiguous park area in the nation, it has villages, towns, farms, dwellings, and businesses as well as parklands that include forests designated by the state as "forever wild." The park agency oversees the quality of the water supplies, and holds rights of approval over new buildings and other expansions of the built environment. Land use is strictly controlled in the interest of preserving the landscape as a park.

Land use controls exist in many countries, and the United States has at various times restored lands devastated by poor management. The US Soil Conservation Service established firm principles of management to combat the central continental dust bowl of the 1920s and 1930s as well as serious problems of erosion of agricultural soils around the entire nation. These efforts, combined with others by the US Department of Agriculture, produced a revolution from the 1920s to the 1940s in land management and massive improvements in water quality.

In a somewhat more narrowly focused effort, New York City relies heavily for its water supply on the Hudson River Watershed north of the city. The extensive drainages that serve the city west of the Hudson are unfiltered, and the city takes special interest in management of the land. When faced earlier with the need for multi-billion-dollar investments in filtering its water supply as land use in the drainage basins intensified, the city decided to invest instead in land use controls in all the critical western basins. (The eastern basin, the Croton River, produces about 10 percent of the total supply and was already being filtered.) The investment in the management of land and drainage basis was expensive, but it saved billions by avoiding the construction and maintenance of a series of filtration plants. The investment also preserved extensive forested landscapes in the region, and brought systematic management of water and wastes to businesses, towns, and residences in the region.[21]

A far-reaching, powerful example for US land and water management might be set immediately by the US government through a series of bold steps entirely within lands federally owned and controlled. The first would entail managing the forests and lands to contribute to the global effort to stabilize the heat-trapping gas content of the atmosphere. That would involve a clear decision to reestablish forests where they have been cut, rebuild the native vegetation where it has been lost to heavy grazing on Bureau of Land Management lands leased to ranchers, and control all harvests to assure that the standing stock of carbon on public lands is rising, not diminishing. It is a simple decision in the context already established as national policy by Congress in ratifying the 1992 FCCC.

There could be no better moment to institute such innovations in the United States in land and water management as a model for the world. Transition in energy sources to renewables is an opportunity not to be missed for it opens doors to the very independence all desire and responsibility for management of local resources that all seek. The major transition lies in the virtually ubiquitous potential of the local production of electricity from solar panels or wind. The end of the fossil-fueled era should be celebrated at every turn as we focus on building local self-sufficiency for energy, food, and environmental stability into the indefinite future.

There are many intermediate steps in making the transition to a renewable energy world. Still lacking is the efficient storage of energy from renewable sources. But that, too, will come. Even now, with electric energy increasingly available locally, water can be split to produce hydrogen, which can be stored, transported, and used either directly as a fuel or source of electricity in a fuel cell. To be sure, the production and use of hydrogen is inefficient and costly by comparison with the current price of oil or gas, but the cost of continuing the fossil-fueled world is virtually infinite: the demise of this civilization in the time of our lives or soon thereafter. If we are to have a future worth having, the fossil-fueled era must come to a close, as must the era of nuclear power and nuclear weapons, and giant industrialized agricultural ventures.

* * *

Today's seven-billion-person-world is far beyond the laissez-faire world of two billion or less in which our major governmental and economic systems were developed. It was a world that seemed biophysically stable, and too large to disrupt even by cataclysmic war.

Until we disrupted it.

Now, when seven people stand where two or less stood only a few decades earlier, there is reason to reconsider the rules governing the sharing of habitat. The problem of management is intensified roughly according to the n-squared law. And if the personal powers of each to command resources have increased in the interim, at least in the industrialized segments of the world, and each commands far more space in the world, the challenges of governance become still more intense. Libertarian arguments that the free market system can or should replace governmental controls collapse totally. Robert Kuttner in an excellent 2014 essay on Karl Polanyi, a twentieth-century critic of capitalism, addressed an earlier contention by economist Friedrich von Hayek that "democratic forms of state planning were bound to end in the totalitarianism of a Stalin or a Hitler." "But," Kuttner continues, "70 years later there is not a single case of social democracy leading to dictatorship, while there are dozens of tragic episodes of market excess destroying democracy."[22]

Even more urgent is responding effectively to the series of scientific insights into not only the evolutionary origins of the biosphere and details of its biotic structure but also evidence of the current trends toward its destruction and the disruption of human habitat, outlined here and in dozens of scholarly reports over more than forty years. The response requires immediate action by government in moving rapidly away from fossil fuels and into renewable energy of all types. Henry M. Paulson, former secretary of the treasury, has recently joined in advocating an immediate tax on fossil fuels to encourage the transition.[23] Such a tax, systematically increasing with time, with proceeds assigned to the rapid development of energy alternatives and encouragement of other nations less capable of making the transition rapidly, is the most direct route. A cap-and-trade system similar to those already in use for carbon emissions in the northeastern United States and California (see chapter 6) could also be effective, if properly managed. Such innovations are the more urgent because action has been so long delayed. At the same time, the rapid development of local land use plans is needed to favor the retention and storage of carbon in terrestrial ecosystems of all latitudes. To give such planning scientific underpinnings, the global scholarly community at all levels, including schools and students of every age, could turn a sharply inquisitive eye on local ecosystems and how to manage them for maximum benefit over the next decades as we make the shift toward the climatic regimes of the late nineteenth and early twentieth centuries. The emphasis would turn to carbon storage on land, retaining existing stocks and rebuilding stocks lost. Judgments need to be made as to how best to use land for agriculture in a new low-carbon world where an emphasis on carbon storage competes with an agriculture focused not on corporate enrichment but rather on local or regional needs for food.[24] The tundra and northern forests as well as grasslands draw special attention because of their vast area and potential for continuing carbon storage. Civilization faces a biophysical crisis born of corporate excesses, "bad manners" that have led to the poisoning of the human habitat, compounded by rapidly increasing human numbers. How did the corporate world manage to be allowed amorality in contributing to destructive growth and environmental corruption? And how did it manage to be taken seriously in its argument before government and the public that it had the right to poison the world?

That "moment" of inattention must give way as we put back into practice an ancient rule fundamental to all civilizations, including our own:

Sic utere tuo ut alienum non laedas.

NOTES

PREFACE

1. See, for example, Paul Hawken, Amory Lovins, and L. Hunter Lovins, *Natural Capitalism: The Next Industrial Revolution* (London: Earthscan, 1999); Lester R. Brown, *World on the Edge: How to Prevent Environmental and Economic Collapse* (New York: W. W. Norton and Company, 2011); James Gustave Speth, *The Bridge at the Edge of the World: Capitalism, the Environment, and Crossing from Crisis to Sustainability* (New Haven, CT: Yale University Press, 2008); James Gustave Speth, *America the Possible: Manifesto for a New Economy* (New Haven, CT: Yale University Press, 2012); Paul R. Ehrlich and Anne H. Ehrlich, *Betrayal of Science and Reason: How Anti-Environment Rhetoric Threatens Our* Future (Washington, DC: Island Press, 1996); John H. Adams and Patricia Adams, *A Force for Nature*, contrib. George Black (New York: Chronicle Books, 2010); "Richard E. Ayres," Ayres Law Group LLP, http://www.ayreslawgroup. com/attorneys/Richard-E.-Ayres (accessed March 20, 2015).

2. Speth, *Bridge at the Edge of the World*, 6.

3. The GDP for the world is the sum of national accounts and is the same as the gross world product. See J. Bradford DeLong, "Estimating World GDP, One Million B.C. to Present" (paper, University of California at Berkeley, 1998).

4. For a lucid account of the somewhat circular evolution of land use law and civil rights in the United States over time from early settlement on, see Eric T. Freyfogle, *The Land We Share: Private Property and the Common Good* (Washington, DC: Island Press, 2003).

5. Garrett Hardin, "The Tragedy of the Commons," *Science* 162, no. 3859 (December 1968): 1243–1248.

CHAPTER 1: IN THE BEGINNING

1. Vladimir I. Vernadsky, *The Biosphere: Complete Annotated Edition* (New York: Copernicus Books, 1998). See also G. Evelyn Hutchingson, "The Biosphere," the

introductory article of the *Scientific American* 223, no. 3 (September 1970), an issue devoted in its entirety to the topic.

2. Daniel J. Hillel, *Out of the Earth: Civilization and the Life of the Soil* (New York: Macmillan, 1991); Daniel Zohary and Maria Hopf, *Domestication of Plants in the Old World: The Origin and Spread of Cultivated Plants in West Asia, Europe, and the Nile Valley* (London: Oxford University Press, 2001); Karl S. Zimmer, "Multilevel Geographies of Seed Networks and Seed Use in Relation to Agrobiodiversity Conservation in the Andean Countries," in *Globalization and New Geographies of Conservation*, ed. Karl S. Zimmer (Chicago: University of Chicago Press, 2006), 141–165.

3. Charles Darwin, *On the Origin of Species: A Facsimile of the First Edition* (Cambridge, MA: Harvard University Press, 1964), 489.

4. George Perkins Marsh, *The Earth as Modified by Human Action: Last Revision of Man and Nature* (New York: Scribner's, 1885), vii.

5. Chris Hedges and Joe Sacco, *Days of Destruction, Days of Revolt* (New York: Perseus, Nation Books, 2012), xiv–x. See also James Gustave Speth, *Angels by the River: A Memoir* (White River Junction, VT: Chelsea Green, 2014).

6. Garrett Hardin, "The Tragedy of the Commons," *Science* 13, no. 162 (December 1968), 1243–1248.

7. James Inhofe, *The Greatest Hoax: How the Global Warming Conspiracy Threatens Your Future* (Washington, DC: WND Books, 2012).

8. Stephen Lacy, "AEI Economist Zycher Makes Head-Exploding Claims about Cost of Renewables," *Climate Progress* (blog), Think Progress, February 28, 2012, http://thinkprogress.org/climate/2012/02/28/430384/aei-economist-zycher-cost-of-renewables (accessed April 21, 2015).

9. Anthony J. McMichael, "Insights from Past Millennia into Climatic Impacts on Human Health and Survival," *Proceedings of the National Academy of Sciences of the United States of America* 109, no. 13 (March 27, 2012): 4730–4737.

10. Lester R. Brown, *Outgrowing the Earth: The Food Security Challenge in an Age of Falling Water Tables and Rising Temperatures* (New York: W. W. Norton and Company, 2004).

11. Thomas Robert Malthus, *An Essay on the Principle of Population* (London: St. Paul's Church-Yard, J. Johnson, 1798).

12. United Nations, Population Division World Population Prospects. 2012. See also "World Population Data Sheet," Population Reference Bureau, Washington, DC, 2013.

13. Hardin, "Tragedy of the Commons."

CHAPTER 2: NUCLEAR ENERGY

1. Scott Kaufman, *Project Plowshare: The Peaceful Use of Nuclear Explosives in Cold War America* (Ithaca, NY: Cornell University Press, 2012).

2. David Bradley, *No Place to Hide 1946/1984* (Hanover, NH: University Press of New England, 1983).

3. Bertil Åberg and Frank P. Hungate, eds., *Radioecological Concentration Processes: Proceedings of an International Symposium Held in Stockholm 25–29 April 1966* (London: Pergamon Press, 1967).

4. "Nuclear Power Plants, Worldwide," European Nuclear Society, January 15, 2015, http://www.euronuclear.org/info/encyclopedia/n/nuclear-power-plant-world-wide. htm (accessed May 29, 2015).

5. Ionizing radiation is high-energy electromagnetic radiation that can break molecules into a charged (ionized) and chemically reactive state. Such molecules can recombine to form dysfunctional forms that interfere with normal life processes. Such breaks in chromosomes are "mutations," virtually all deleterious.

6. Arnold H. Sparrow, "Cytological Changes Induced by Ionizing Radiations and Their Possible Relation in the Production of Useful Mutations in Plants," in *Work Conference on Radiation Induced Mutations,Biology Department,* ed. A. H. Sparrow (Upton, NY: Brookhaven National Laboratory, 1956), 76–113.

7. Ian Hore-Lacy and World Nuclear Association, "Price-Anderson Act of 1957, United States," Encyclopedia of Earth, December 7, 2009, http://www.eoearth.org/ view/article/155347 (accessed May 29, 2015).

8. Sparrow, "Cytological Changes Induced by Ionizing Radiations," 76–113.

9. Eugene Odum, *Fundamentals of Ecology* (Philadelphia: Saunders,1953).

10. E. Lucy Braun, *Deciduous Forests of Eastern North America* (New York: Hafner Publishing Company, 1950). Braun described the postglacial history of this major North American forest in detail in her book, now famous among botanists and ecologists.

11. George M. Woodwell and E. Cuyler Hammond, *A Descriptive Technique for Study of the Effects of Chronic Ionizing Radiation on a Forest Ecological System* (Upton, NY: Brookhaven National Laboratory, 1962).

12. George M. Woodwell, "Design of the Brookhaven Experiment on the Effects of Ionizing Radiation on a Terrestrial Ecosystem," *Radiation Botany* 3, no. 2 (1963): 125–133.

13. George M. Woodwell, "Effects of Ionizing Radiation on Terrestrial Ecosystems," *Science* 138, no. 3540 (November 2, 1962): 572–577.

14. George M. Woodwell, *The Earth in Transition: Patterns and Processes of Biotic Impoverishment* (New York: Cambridge University Press, 1991).

15. Eville Gorham and Alan G. Gordon, "The Influence of Smelter Fumes upon the Chemical Composition of Lake Waters near Sudbury, Ontario, and upon the Surrounding Vegetation," *Canadian Journal of Botany* 38, no. 4 (July 1960): 477–487.

16. S. A. Cain, *Foundations of Plant Geography* (New York: Harper and Brothers Publishers, 1944).

17. Braun, *Deciduous Forests of Eastern North America*; H. J. Oosting, *The Study of Plant Communities* (San Francisco: W. H. Freeman, 1958).

18. Frederic E. Clements, *Plant Succession: An Analysis of the Development of Vegetation* (Washington, DC: Carnegie Institute of Washington, 1916), 242; Frederic E. Clements, "Nature and Structure of the Climax," *Journal of Ecology* 24, no. 1 (1936): 252–284.

19. George M. Woodwell, "The Earth under Stress: A Transition to Climatic Instability Raises Questions about Patterns of Impoverishment," in *The Earth in Transition: Patterns and Processes of Biotic Impoverishment*, ed. George M. Woodwell (New York: Cambridge University Press, 1990), 3–8.A well-founded and carefully observed principle of horticulture and forestry requires consideration of the source of any plantation in recognition of the quite-specific adaptation of all organisms to their place of origin. Is the organism likely to survive in the new circumstance? For a discussion, see Oosting, *The Study of Plant Communities*, 25–28.See also F. Stuart Chapin III, Pamela A. Matson, and Peter Vitousek, *Principles of Terrestrial Ecosystem Ecology* (New York: Springer, 2012).

20. John W. Goffman, *Radiation and Human Health* (San Francisco: Sierra Club Book, 1981).

21. Hore-Lacy and World Nuclear Association, "Price-Anderson Act of 1957."

22. "Chernobyl Accident 1986," World Nuclear Association, 2014, http://www.world-nuclear.org/info/safety-and-security/safety-of-plants/chernobyl-accident (accessed May 30, 2015). The World Nuclear Association is the professional association of specialists.

23. Nikolaus Riehl and Frederick Seitz, *Stalin's Captive: Nikolaus Riehl and the Soviet Race for the Bomb* (Philadelphia: American Chemical Society, 1996); Zhores A. Medvedev, *A Nuclear Disaster in the Urals* (New York: W. W. Norton, 1980).

24. See http://www.geocurrents.info/place/russia-ukraine-and-caucasus/siberia/kyshtym-57-a-siberian-nuclear-disaster (accessed May 30, 2015).

25. *On the Fukushima Nuclear Disaster before the Joint Hearings of the Subcommittee on Clean Air and Nuclear Safety and the Committee on Environment and Public Works*, 112th Cong. (2011) (statement of Thomas B. Cochran, PhD, senior scientist, Nuclear

Program, Natural Resources Defense Council, Inc.), https://www.nrdc.org/nuclear/ files/tcochran_110412.pdf (accessed May 30, 2015).

26. "Nuclear Power Plants Worldwide."

27. Robert Alvarez, *Spent Nuclear Fuel Pools in the U.S.: Reducing the Deadly Risks of Storage* (Washington, DC: Institute for Policy Studies, 2011).

28. Mark Schrope, "Nuclear Power Prevents More Deaths Than It Causes," *Chemical and Engineering News*, April 2, 2013, https://cen.acs.org/articles/91/web/2013/04/ Nuclear-Power-Prevents-Deaths-Causes.html (accessed May 30, 2015); Pushker A. Kharecha and James E. Hansen, "Prevented Mortality and Greenhouse Gas Emissions from Historical and Projected Nuclear Power," *Environmental Science and Technology* 47, no. 9 (2013): 4889–4895.

29. "Nuclear Facts," Natural Resources Defense Council, 2007, http://www.nrdc.org/ nuclear/plants/plants.pdf (accessed May 30, 2015).

30. Eric Schlosser, *Command and Control: Nuclear Weapons, the Damascus Accident, and the Illusion of Safety* (New York: Penguin Books, 2013).

31. Hannes Alfven, "Energy and Environment," *Bulletin of the Atomic Scientists* 29, no. 5 (1972): 5–8.

32. Sandra Steingraber, *Living Downstream: An Ecologist's Personal Investigation of Cancer and the Environment* (Boston: Da Capo Press, 2010).

33. Adam J. White, "Yucca Mountain: A Post-Mortem," *New Atlantis* 37 (2012): 3–19.

CHAPTER 3: DDT DRIVES A GEOCHEMICAL TEMPEST

1. George M. Woodwell, Charles F. Wurster, and Peter A. Isaacson, "DDT Residues in an East Coast Estuary: A Case of Biological Concentration of a Persistent Insecticide," *Science* 156, no. 3776 (May 1967): 821–824.

2. George M. Woodwell, and Frederick T. Martin, "Persistence of DDT in Soils of Heavily Sprayed Forest Stands," *Science* 145, no. 3631 (July 1964): 481–483.

3. P. A. Butler, "Residues in Fish, Wildlife, and Estuaries: Organochlorine Residues in Estuarine Mollusks," *Pesticides Monitoring Journal* 6, no. 4 (March 1973).

4. Charles F. Wurster treats the human toxicology of DDT and other chlorinated hydrocarbons extensively in *DDT Wars: Rescuing Our National Bird, Preventing Cancer, and Creating the Environmental Defense Fund* (New York: Oxford University Press, 2015). Research experience with small mammals led to the conclusion, widely held by scientists, that "DDT was a cancer danger to humans" (106–107).

5. DDT was highly popular in agriculture but despite its popularity, turned out not to be essential. During this period of the early 1960s, for several years I was a member

of the New York State Pesticides Control Board, which was specifically charged with recommending how to manage pesticides. I recall the testimony of the NY state commissioner of agriculture to the board to the effect that DDT was absolutely essential to agriculture in New York and its use could not be curtailed. Within weeks the governor at the time, Nelson Rockefeller, had decided that DDT had to be controlled, and announced a policy and program. The commissioner suddenly discovered that removing DDT would have "no effect" on agricultural productivity and unabashedly announced it to the board.

6. See for example, the warning, "Avoid spraying water. Because of the sensitivity of fishes and crabs to DDT, do not apply directly to streams, lakes and coastal bays." Clarence Cottam and Elmer Higgins, "DDT and Its Effect on Fish and Wildlife," *Journal of Economic Entomology* 39, no. 1 (February 1946): 44.

7. See, for example, Woodwell and Martin, "Persistence of DDT in Soils of Heavily Sprayed Forest Stands," 481–483. See also A. D. Michael, C. R. Van Raalte, and L. S. Brown, "Long-Term Effects of an Oil Spill at West Falmouth, Massachusetts," *International Oil Spill Conference Proceedings* 1975, no. 1 (March 1975): 573–582.

8. This problem became more acute as research budgets at universities dropped as the more conservative interests in Congress gained sway. The pressures on research budgets pushed scientists to seek money from industries for agricultural research—a turn that gave industries even greater control over the flow of activities in the agricultural universities.

9. Wurster, *DDT Wars*.

10. Jennifer Sass, "Neonicotinoid Pesticides—Bad for Bees, and May Be Bad for People Too," *Switchboard* (blog), Natural Resources Defense Council, September 17, 2014, http://switchboard.nrdc.org/blogs/jsass/neonicotinoid_pesticides_-_bad.html (accessed June 1, 2015). See also "What You Should Know about 2, 4-D," Natural Resources Defense Council, http://www.nrdc.org/health/pesticides/2-4-d.asp (accessed March 8, 2012). See also Francisco Sánchez-Bayo, "The Trouble with Neonicotinoids," *Science* 346, no. 6211 (November 2014): 806–807.

11. Gennaro Di Prisco, Valeria Cavaliere, Desiderato Annoscia, Paola Varricchio, Emilio Caprio, Francesco Nazzi, Giuseppe Gargiulo, and Francesco Pennacchio, "Neonicotinoid Clothianidin Adversely Affects Insect Immunity and Promotes Replication of a Viral Pathogen in Honey Bees," *Proceedings of the National Academy of Sciences* 110, no. 46 (August 2013); Damian Carrington, "US Government Sued over Use of Pesticides Linked to Bee Harm," *Guardian*, March 22, 2013. See also Caspar A. Hallmann, Ruud P. B. Foppen, Chris A. M. van Turnhout, Hans de Kroon, and Eelke Jongejans, "Declines in Insectivorous Birds Are Associated with High Neonicotinoid Concentrations," *Nature* (July 2014). For more on the neurological effects, see Motohiro Tomizawa and John E. Casida, "Selective Toxicity of Neonicotinoids Attributable to Specificity of Insect and Mammalian Nicotinic Receptors," *Annual Review of Entomology* 48 (June 2002): 339–364.

12. There is a rich background of research on neonicotinoids and their diversity of biophysical characteristics. The most powerful summary of their hazards is that offered by NRDC staff in a recent appeal to the EPA to ban neonicotinoids because of the accumulating evidence that they are responsible for the destruction of bee colonies around the world. See Margaret Hsieh and Jennifer Sass, Environmental Protection Agency, "Petition for Interim Administrative Review of Neonicotinoid Pesticides," petition, July 7, 2014, http://docs.nrdc.org/health/files/hea_14070701a. pdf (accessed June 1, 2015).

13. George H. Hallberg, "From Hoes to Herbicides: Agriculture and Groundwater Quality," *Journal of Soil and Water Conservation* 41, no. 6 (November–December 1986): 357–364; William A. Battaglin, Edward T. Furlong, and Michael R. Burkhardt, *Concentration of Selected Sulfonylurea, Sulfonamide, and Imidazolinone Herbicides, Other Pesticides, and Nutrients in 71 Streams, 5 Reservoir Outflows, and 25 Wells, in the Mid-western United States* (Denver, CO: US Geological Survey, 1998). See also Patricia L. Toccalino, Robert J. Gilliom, Bruce D. Lindsey, and Michael G. Rupert, "Pesticides in Groundwater of the United States: Decadal-Scale Changes, 1993–2011," *Groundwater* 52, no. S1 (September 2014): 112–125.

14. International Survey of Herbicide Resistant Weeds, www.weedscience.com (accessed March 20, 2015).

15. Cheryl Lyn Dybas, "Polar Bears Are in Trouble—and Ice Melt's Not the Half of It," *BioScience* 62, no. 12 (2012): 1014–1018. See also J. H. Park, A. H. Goldstein, J. Tomovsky, S. Fares, R. Weber, J. Karlik, and R. Holzinger, "Active Atmosphere-Eco-system Exchange of the Vast Majority of the Detected Volatile Organic Compounds," *Science* 341, no. 6146, (August 2013): 643–647.

16. Theo Colborn, Dianne Dumanoski, and John Peter Myers, *Our Stolen Future: Are We Threatening Our Fertility, Intelligence, and Survival?—A Scientific Detective Story* (New York: Plume, 1996).

17. Frederica P. Perera, Shuang Wang, Virginia Rauh, Hui Zhou, Laura Stigter, David Camann, Wieslaw Jedrychowski, Elzbieta Mroz, and Renata Majewska, "Prenatal Exposure to Air Pollution, Maternal Psychological Distress, and Child Behavior," *Pediatrics* 132, no. 5 (November 2013): 284–294.

18. Linda A. Deegan, Jennifer L. Bowen, Deanne Drake, John W. Fleeger, Carl T. Friedrichs, Kari A. Galván, John E. Hobbie et al., "Susceptibility of Salt Marshes to Nutrient Enrichment and Predator Removal," *Ecological Applications* 17 (2007): S42–S63.

19. Jerry H. Yen, "Chemical Regulation in the European Union (EU): Registration, Evaluation, and Authorization of Chemicals," Congressional Research Service, October 23, 2013.

20. Chris Hedges and Joe Sacco, *Days of Destruction, Days of Revolt* (New York: Nation Books, 2012).

21. Peter Lehner and Bob Deans, *In Deep Water: The Anatomy of a Disaster, the Fate of the Gulf, and How to End Our Oil Addiction* (New York: OR Books, 2010).

22. Benjamin Landy, "Meet 'Dirty Dozen' Tax Break #2: The 'Percentage Depletion' Deduction," *Blog of the Century* (blog), Century Foundation, December 10, 2012, http://tcf.org/blog/detail/meet-dirty-dozen-tax-break-2-the-percentage-depletion-deduction (accessed June 1, 2015).

23. Dave Hawkins, "ExxonMobil to Humanity: 'Drop Dead (But Keep Your Motors Running,'" *Switchboard* (blog), Natural Resources Defense Council, April 3, 2014, http://switchboard.nrdc.org/blogs/dhawkins/exxonmobil_to_humanity_drop_de. html (accessed June 1, 2015).

24. John W. Goffman, *Radiation and Human Health* (San Francisco: Sierra Club Book, 1981).

CHAPTER 4: CARBON AND THE CLIMATIC DISRUPTION

1. Daniel Nepstad, "Climate Change and the Forest," *American Prospect* 18, no. 8 (August 2007); Daniel Nepstad, *The Amazon's Vicious Cycles* (Gland, Switzerland: World Wildlife Fund International, 2007); "A Report for the United Nations Framework Convention on Climate Change (UNFCCC) Conference of the Parties (COP)," Thirteenth Session, December 3–14, 2007, Bali, Indonesia, http://unfccc.int/meetings/bali_dec_2007/meeting/6319/php/view/reports.php (accessed June 1, 2015); Daniel Nepstad, I. M. Tohver, D. Ray, P. Moutinho, and G. Cardinot."Mortality of Large Trees and Lianas Following Experimental Drought in an Amazon Forest," *Ecology* 88, no. 9 (2007): 2259–2269; Daniel Nepstad et al., *Manejo e Recuperação de Mata Ciliar em Regiões Florestais da Amazônia* [Management and recuperation of riparian zone forests of Amazon forest regions] (Belém, Brazil: Instituto de Pesquisa Ambiental da Amazônia, 2007); Daniel Nepstad, B. Soares-Filho, F. Merry, P. Moutinho, A. Rodrigues, S. Schwartzman, O. Almeida, and S. Rivero, *Reducing Emissions from Deforestation and Forest Degradation (REDD): The Costs and Benefits of Reducing Carbon Emissions from Deforestation and Forest Degradation in the Brazilian Amazon* (Falmouth, MA: Woods Hole Research Center, 2007).

2. Eneas Salati, A. Dall'olio, E. Matsui, J.R. Gat., "Recycling of Water in the Amazon Basin: An Isotopic Study," *Water Resources Research* 15, no. 5 (1979): 1250–1258; Marcello Z. Moreira, Leonel Sternberg, Luiz A. Martinelli, Reynaldo L. Victoria, Edelcilio M. Barbosa, Luiz C. M. Bonates, and Daniel C. Nepstad, "Contribution of Transpiration to Forest Ambient Vapour Based on Isotopic Measurements," *Global Change Biology* 3, no. 5 (October 1997): 439–450.

3. IPCC Fifth Assessment Report (AR5) of the Intergovernmental Panel on Climate Change. Reports of the Intergovernmental Panel on Climate Change (IPCC) are available from Cambridge University Press and the IPCC: https://www.ipcc.ch/report/ar5 (accessed August 19, 2015). See also American Association for the Advancement of Science: Science Panel: "What We Know: The Reality, Risks, and Response to Climate Change" (Washington, DC: American Association for the Advancement of Science, March, 2014).

4. E. W. Mayer, D. R. Blake, S. C. Tyler, Yoshikhiro Makide, D. C. Montague, and F. S. Rowland, "Methane: Interhemispheric Concentration Gradient and Atmospheric Residence Time," *Proceedings of the National Academy of Sciences USA* 79 (1982): 1366–1370.

5. Harriet Sherwood, Helena Smith, Lizzy Davies, and Harriet Grant, "Europe Faces 'Colossal Humanitarian Catastrophe' of Refugees Dying at Sea," *Guardian*, June 2, 2014.

6. Albert Oppog-Ansah, "Climate Makes Refugees Out of Young Ghanaians," Inter Press Service, December 13, 2013.

7. Oeschger and his colleagues at the University of Bern developed techniques for determining the carbon dioxide content of air trapped in the annual layers of glacial ice. In the 1980s, they were ultimately able to produce a compelling record of the history of carbon dioxide and temperature in the atmosphere over eight hundred thousand years, and showed that over that period the concentration of carbon dioxide had not exceeded the approximately 0.280 ppm of 1880. It was an extraordinary advance long sought by scientists, who previously had no basis for discovering the history of the atmosphere before direct measurements started in the mid-1800s. See *Complete Dictionary of Scientific Biography*, s.v. "Hans Oeschger."

8. International Institute for Sustainable Development, *Final Report of the World Commission on Forests and Sustainable Development* (Cambridge: Cambridge University Press, 1999); Nikos Alexandratos and Jelle Bruinsma, "World Agriculture towards 2030–2050" (working paper, Food and Agriculture Division of the United Nations, Global Perspective Studies Team, 2012).

9. George M. Woodwell and Winston. R. Dykeman, "Respiration of a Forest by Carbon Dioxide Accumulation During Temperature Inversions," *Science* 154, no. 3752 (1966): 1031–1034. See also S. C. Wofsy et al., "Net Exchange of CO2 in a Mid-Latitude Forest," *Science* 260, no. 5112 (1993): 1314–1315.

10. Wallace S. Broecker and Tsung Hung Peng, *Greenhouse Puzzles* (Palisades, NY: Eldigio Press, 1998).

11. In fact, by far the largest pool of biotically controlled carbon lies in the boreal forest, its soils, and the deep peat of the tundra. Those carbon deposits are in various forms at middle to high northern latitudes around the world. Our appraisals in

those years of the 1970s into the 1990s suggested that the total carbon in forests and soils globally was three times or more the amount then in the atmosphere. The stability of that carbon is of considerable importance if we are to avoid a serious temperature excursion. See George M. Woodwell and Fred T. Mackenzie, eds., *Biotic Feedbacks in the Global Climatic System* (Oxford: Oxford University Press, 1995).

12. US Department of the Interior, *USGS World Petroleum Assessment 2000* (Washington, DC: US Department of the Interior, June 2003); Gordon J. MacDonald, "Role of Methane Clathrates in Past and Future Climates," *Climatic Change* 16, no. 3 (June 1990): 247–281; Bruce Buffett and David Archer, "Global Inventory of Methane Clathrate: Sensitivity to Changes in the Deep Ocean," *Earth and Planetary Science Letters* 227 (November 2004): 185–199.

13. John Firor and Judith E. Jacobsen, *The Crowded Greenhouse* (New Haven, CT: Yale University Press, 2002). See also William W. Kellogg and Robert Schware, *Climate Change and Society* (Boulder, CO: Westview Press, 1981).

14. Richard S. Williams Jr. and Jane G. Ferrigno, eds., "Satellite Image Atlas of Glaciers of the World" (professional paper 1386, US Geological Survey, 2012).

15. The public welfare is in the custody in most democratic societies of elected representatives from among the public. Extending those responsibilities to environmental issues has been an innovation in the eyes of many, but has moved far over the past three decades from earlier stirrings such as those set before the Commission of the European Communities in 1989. For a discussion, see P. Bourdeau, P. M. Fasella, and A. Teller, eds., *Environmental Ethics: Man's Relationship with Nature-Interactions with Science* (Luxembourg: Commission on the European Communities, 1989). See also Naomi Klein, *This Changes Everything* (New York: Simon and Shuster, 2014); Naomi Oreskes and Erik M. Conway, *The Collapse of Western Civilization* (New York: Columbia University Press, 2014).

16. National Aeronautics and Space Administration (NASA), "2014 Arctic Sea Ice Minimum Sixth Lowest on Record," news release, September 22, 2014, http://www.nasa.gov/press/2014/september/2014-arctic-sea-ice-minimum-sixth-lowest-on-record/#.VL0SL9LF_Y9 (accessed June 2, 2015).

17. Natural Resources Defense Council, "New Orleans, Louisiana: Identifying and Becoming More Resilient to Impacts of Climate Change," July 2011, http://www.nrdc.org/water/files/ClimateWaterFS_NewOrleansLA.pdf (accessed June 2, 2015).

18. NASA's Goddard Space Flight Center, "Typhoon Haiyan," news release, November 20, 2013, http://www.nasa.gov/content/goddard/haiyan-northwestern-pacific-ocean/#.VL0ivtLF_Y8 (accessed June 2, 2015). BBC News, "Typhoon Haiyan: Rebuilding Lives in Tacloban," news release, November 22, 2013, http://www.bbc.com/news/world-asia-25046721 (accessed June 2, 2015).

19. See Williams and Ferrigno, "Satellite Image Atlas."

20. Bill McKibben, "Global Warming's Terrifying New Math," *Rolling Stone* 1162 (August 2012).

21. John Shepherd, "Geoengineering the Climate: An Overview and Update," *Philosophical Transactions of the Royal Society* 370, 4166–4175 (2012); Bert Metz et al., eds., *Carbon Dioxide Capture and Storage* (New York: Cambridge University Press, 2005); Clive Hamilton, *Earthmasters* (New Haven, CT: Yale University Press, 2013); National Academy of Sciences, "Geoengineering Climate: Technical Evaluation of Selected Approaches," news release, February 10, 2015, https://nas-sites.org/americasclimatechoices/studies-in-progress/geoengineering-technical-evaluation-of-selected-approaches (accessed June 2, 2015).

CHAPTER 5: THE GLOBAL COMMONS

1. E. Cuyler Hammond and Daniel Horn, "Smoking and Death Rates: Report on Forty-Four Months of Follow-up of 187,783 Men," *Journal of American Medical Association* 166, no. 11 (March 1958): 1294–1308. See also Richard Doll and Austin Bradford Hill, "Mortality in Relation to Smoking: Ten Years' Observations of British Doctors," *British Medical Journal* 1, no. 5396 (June 1964): 1460–1467.

2. Naomi Oreskes and Erik M. Conway, *Merchants of Doubt* (New York: Bloomsbury Press, 2010); Robert N. Proctor, *Golden Holocaust* (Berkeley: University of California Press, 2011).

3. Hammond and Horn, "Smoking and Death Rates." Op cit

4. Garrett Hardin, "The Tragedy of the Commons," *Science* 13, no. 162 (December 1968), 1243–1248.

5. David W. Orr, *Earth in Mind* (Washington, DC: Island Press, 2004).

6. James Gustave Speth, *America the Possible* (New Haven, CT: Yale University Press, 2012); Al Gore, *The Future* (New York: Random House, 2013).

7. Margaret Sullivan, "After Changes, How Green Is TheTimes?" Public Editor, *New York Times*, November 23, 2013, http://www.nytimes.com/2013/11/24/public-editor/after-changes-how-green-is-the-times.html (accessed June 3, 2015).

8. I remember well a discussion with the heads of *Scientific American* when it was independently owned. The magazine was highly respected by the scientific community, who thought of the periodical as their own, and ascribed "responsibilities" in representing science fairly and accurately. The responsibility to accuracy was never questioned. But the choice of topics was, and the publishers' response was a quite-candid denial of any responsibility beyond selling copies and financial success. Even that ultimately dimmed, and the magazine was sold to a German publishing enterprise.

9. James M. Acton and Mark Hibbs, *Why Fukushima Was Preventable* (Washington, DC: Carnegie Endowment for International Peace, 2012).

10. On the Chevron-Ecuador saga, see Public Citizen, "Ecuador's Highest Court vs. a Foreign Tribunal," December 11, 2013, http://citizen.typepad.com/eyesontrade/2013/12/ecuadors-highest-court-vs-a-foreign-tribunal-who-will-have-the-final-say-on-whether-chevron-will-pay.html (accessed June 3, 2015); Friends Committee on National Legislation, "Exposing Biggest Trade Deal in U.S. History," news release, January 9, 2014, http://fcnl.org/issues/energy/exposing_biggest_trade_deal (accessed June 3, 2015).

11. An area extending twenty-five miles to the northwest of the reactor site was contaminated with long-lived radioactive fallout high in Cs 137, a gamma emitter with a thirty-year half-life. See Mark Holt, Richard J. Campbell, and Mary Beth Nikitin, "Fukushima Nuclear Disaster," Congressional Research Service, January 18, 2012; Alex Rosen, "Effects of the Fukushima Nuclear Meltdown on Environment and Health," University Clinic Dusseldorf, March 9, 2012.

12. P. Bailly du Bois, P. Laguionie, D. Boust, I. Korsakissok, D. Didier, and B. Fiévet. "Estimation of Marine Source Term Following Fukushima Daiichi Accident," *Journal of Environmental Radioactivity* 114 (December 2012): 2–9. See also Patrick J. Kiger, "Fukushima Radioactive Water Leak: What You Should Know," *National Geographic News*, August 9, 2013.

13. Steve Forbes, personal letter to the author, September 11, 2007.

14. "Technology Assessment and Congress," Office of Technology Assessment Archives provided by the Federation of American Scientists, http://ota.fas.org/technology_assessment_and_congress (accessed April 3, 2015).

15. Mike Glover, "Newt Gingrich: EPA Should Be Eliminated," *Huffington Post*, January 25, 2011.

16. Associated Press, "2 Paint Companies Lose Big Calif. Case over Lead Paint," CBS News, December 17, 2013; Associated Press, "GM Didn't Warn Customers about Ignition Switch Problems for 11 Years," *Huffington Post*, July 19, 2014.

17. Elisa Wood, "Three Years after Fukushima: Lulled into the Myth of Safety?" www.renewableenergyworld.com (accessed March 11, 2014).

18. Peter Schwartz and Doug Randall, "An Abrupt Climate Change Scenario and Its Implications for United States National Security," Climate Institute, October 2003.

19. IPCC, 2007: Summary for Policymakers, in *Climate Change 2007: The Physical Science Basis. Contribution of Working Group I to the Fourth Assessment Report of the Intergovernmental Panel on Climate Change*, ed. S. Solomon, D. Qin, M. Manning, Z. Chen, M. Marquis, K. B. Averyt, M. Tignor, and H. L. Miller (Cambridge: Cambridge University Press, 2007). See also Michael T. Coe and Jonathan A. Foley, "Human and

Natural Impacts on the Water Resources of the Lake Chad Basin," *Journal of Geophysical Research: Atmospheres* 106, no. D4 (February 2001): 3349–3356.

20. Population Reference Bureau, "2013 Population Reference Sheet," Washington, DC.

21. Paul Adams, "Migration Surge Hits EU as Thousands Flock to Italy," BBC News, May 30, 2014, http://www.bbc.com/news/world-europe-27628416 (accessed June 3, 2015).

22. Eric Lipton and Clifford Krauss, "Fossil Fuel Industry and Ads Dominate TV Campaign," *New York Times*, September 13, 2012.

23. Danielle Droitsch, "Prominent Canadians Visit Washington DC to Share the Truth about Canada's Record," *Switchboard* (blog), Natural Resources Defense Council, October 11, 2013.

24. Allan Woods, "Conservative Government Shutting Down World-Class Freshwater Research Facility in Northern Ontario," *Toronto Star*, May 17, 2012.

25. Clifford Krauss, "Plan to Ban Oil Drilling in the Amazon Is Dropped," *New York Times*, August 17, 2013, B1.

26. Sheldon Wolin, "Inverted Totalitarianism," *Nation*, May 1, 2003.

27. Lester Brown, *Plan B 4.0* (New York: W. W. Norton, 2009), 18–23. See also Fund for Peace, "The Failed States Index," *Foreign Policy* (July–August 2005).

28. Lester Brown, *State of the World 1995* (Washington, DC: Worldwatch Institute, 1995).

29. Paul Farmer, *Haiti after the Earthquake* (New York: Perseus, 2011).

30. George Perkins Marsh, *The Earth as Modified by Human Action* (1864; repr., New York: C. Scribner's Sons, 1885).

31. Steven Pearlstein, "When Shareholder Capitalism Came to Town," *American Prospect* 25, no. 2 (March 2014): 40–54.

32. Herman E. Daly, *Beyond Growth: The Economics of Sustainable Development* (Boston: Beacon Press, 1996), 8.

33. Herman E. Daly, "From Empty-World Economics to Full-World Economics: Recognizing an Historical Turning Point in Economic Development," in *World Forests for the Future: Their Use and Conservation*, ed. Kilaparti Ramakrishna and George M. Woodwell (New Haven, CT: Yale University Press, 1993), 79–91.

34. In lectures and discussions, Speth suggests the need for a "New Economy" that embraces much of what is suggested here with respect to expanded responsibilities of corporate interests. I believe a much larger formal role for science and environment is appropriate, as explored later in this book. See Speth, *America the Possible*.

CHAPTER 6: CLIMATE IN THE TIDES OF POLITICS

1. Parker F. Jones, "Goldsboro Revisited: Account of Hydrogen Bomb Near-Disaster over North Carolina—Declassified," October 22, 1969, http://www.theguardian.com/world/interactive/2013/sep/20/goldsboro-revisited-declassified-document (accessed June 4, 2015). The document, obtained by the investigative journalist Eric Schlosser under the Freedom of Information Act, gives the first conclusive evidence that the United States was narrowly spared a disaster of monumental proportions when two Mark 39 hydrogen bombs were accidentally dropped over Goldsboro, North Carolina, on January 23, 1961. The bombs fell to earth after a B-52 bomber broke up in midair, and one of the devices behaved precisely as a nuclear weapon was designed to do in warfare: its parachute opened, its trigger mechanisms engaged, and only one low-voltage switch prevented untold carnage. Each bomb carried a payload of four megatons—the equivalent of four million tons of TNT explosive. Had the device detonated, lethal fallout could have been deposited over Washington, Baltimore, Philadelphia, and as far north as New York City—putting millions of lives at risk. See National Public Radio's *Living on Earth*, September 21, 2013.

2. Daily Climate, Twitter post, February 8, 2015, 7:00 a.m., http://twitter.com/thedailyclimate/status/564438619864047616 (accessed June 4, 2015).

3. George M. Woodwell, Gordon J. MacDonald, Roger Revelle, and C. David Keeling, "A Report to the Council on Environmental Quality, Executive Office of the President," July 1979.

4. Jonathan Lash, *A Season of Spoils: The Reagan Administration's Attack on the Environment* (New York: Pantheon, 1984). Lash, a distinguished environmental lawyer at one time on the NRDC staff, showed the ways in which public resources were being squandered on private interests during the Reagan era.

5. "Rapid Global Warming: Worse with Neglect: Testimony of G. M. Woodwell before the Committee on Energy and Natural Resources," US Senate, June 23, 1988,

6. "Biotic Influences and Global Warming: Cause and Effect: Testimony before the Subcommittee on Water Resources of the Committee on Interior and Insular Affairs," US House of Representatives, June 1988.

7. Elizabeth Douglass, "Deepening Ties between Exxon and Russia Run Counter to U.S. Efforts to Punish Putin," *InsideClimate News* (Brooklyn), August 27, 2014.

8. Michael Wines, "Wells Dry, Fertile Plains Turn to Dust," *New York Times*, May 19, 2013.

9. Suzanne Goldenberg, "Conservative Groups Spend up to $1bn a Year to Fight Action on Climate Change," *Guardian* (London), December 20, 2013, http://www.theguardian.com/environment/2013/dec/20/conservative-groups-1bn-against-climate-change (accessed April 2015).

10. James Inhofe, *The Greatest Hoax: How the Global Warming Conspiracy Threatens Your Future* (Washington, DC: WND Books, 2012). See also Joseph Romm, Think Progress, http://thinkprogress.org/climate/2015/03/04/3629466/snowball-inhofe-meet-the-press-climate (accessed October 1, 2015).

11. Brendan Demelle, "Penn State Completely Exonerates Climate Scientist Michael Mann on Bogus Climategate Accusations," *Desmog* (blog), July 1, 2010, http://www.desmogblog.com/penn-state-completely-exonerates-climate-scientist-michael-mann-bogus-climategate-accusations (accessed April 2015). See also Michael Mann, *The Hockey Stick and the Climate Wars: Dispatches from the Front Lines* (New York: Columbia University Press, 2012).

12. Bruce Mohn, "William Koch: Delay, Delay, Delay," *CommonWealth Magazine* 11, April 9, 2013.

CHAPTER 7: THE ADAPTATION MYTH

1. This issue has long been of great interest and widely discussed in scientific circles. A large, recent federal grant has bought the physical oceanographers together to use new equipment and techniques to examine the topic in detail.

2. See "Typhoon Haiyan (Yolanda), USAID, http://www.usaid.gov/haiyan (accessed June 6, 2015).

3. See chap. 4. See also Hans Oeschger and C. C. Langway, eds., The Environmental Record in Glaciers and Ice Sheets (New York: Wiley, 1989); Dieter Lüthi, Martine Le Floch, Bernhard Bereiter, Thomas Blunier, Jean-Marc Barnola, Urs Siegenthaler, Dominique Raynaud, Jean Jouzel, Hubertus Fischer, Kenji Kawamura, and Thomas F. Stocker, "High-resolution Carbon Dioxide Concentration Record 650,000–800,000 Years before Present." *Nature* 453 (May 15, 2008): 379–382.

4. "Earth Is Cooling ... No It's Warming," *Earth Observatory*, http://earthobservatory.nasa.gov/Features/GISSTemperature/giss_temperature2.php (accessed April 2015). See also National Research Council, *Abrupt Impacts of Climate Change: Anticipating Surprises* (Washington DC: National Academies Press, 2013); AAAS Climate Science Panel, *What We Know: The Reality, Risks, and Response to Climate Change* (Washington, DC: American Association for the Advancement of Science, 2014).

5. James Hansen, Pushker Kharecha, Makiko Sato, Valerie Masson-Delmotte, Frank Ackerman, David J. Beerling, Paul J. Hearty et al., "Assessing 'Dangerous Climate Change': Required Reduction of Carbon Emissions to Protect Young People, Future Generations, and Nature," *PLOS One*, December 3, 2013.

6. Suzanne Goldenberg, "Conservative Groups Spend up to $1bn a Year to Fight Action on Climate Change," *Guardian* (London), December 20, 2013, http://www.theguardian.com/environment/2013/dec/20/conservative-groups-1bn-against-climate-change (accessed April 2015).

7. John P. Holdren, "Climate-Change Science and Policy: What Do We Know? What Should We Do?" (paper presented at the Kavli Prize Symposium, Oslo, September 6, 2010).

8. EPA Press Office, "EPA Proposes First Guidelines to Cut Carbon Pollution from Existing Power Plants; Clean Power Plan Is Flexible Proposal to Ensure a Healthier Environment, Spur Innovation, and Strengthen the Economy," Washington, DC, June 2, 2014.

9. IPCC, *Climate Change 2014: Mitigation of Climate Change* (Geneva, Switzerland: Intergovernmental Panel on Climate Change, 2014).

10. Quoted in Michael Babad, "Exxon Mobil CEO: 'What Good Is It to Save the Planet If Humanity Suffers?'" *Globe and Mail* (Toronto), May 30, 2013.

11. Quoted in Bruce Mohl, "Look Who's Talking," *CommonWealth Magazine* (Spring 2013).

12. Reports of the National Snow and Ice Data Center, http://nsidc.org/research (accessed April 2015). See also "Arctic Sea Ice Decline," *Weather Underground* (blog), http://www.wunderground.com/climate/SeaIce.asp (accessed April 2015).

13. Fred T. Mackenzie and Andreas J. Andersson, "The Marine Carbon System and Ocean Acidification during Phanerozoic Time," *Geochemical Perspectives* 2, no. 1 (April 2013): 1–227.

14. William H. Schlesinger, "Soil Respiration and Changes in Soil Carbon Stocks," in *Biotic Feedbacks in the Global Climatic System: Will the Warming Feed the Warming?*, ed. George M. Woodwell and Fred T. Mackenzie (Oxford: Oxford University Press, 1995). See also Richard T. Conant Michael, Ryan, Goran I Agren, Hannah Birge, and Eric Davidson, "Temperature and Soil Organic Matter Decomposition Rates—Synthesis of Current Knowledge and a Way Forward," *Global Change Biology* 17, no. 11 (November 2011): 3392–3404.

15. Eville Gorham, "The Biogeochemistry of Northern Pleatlands and Its Possible Responses to Global Warming," in *Biotic Feedbacks in the Global Climatic System: Will the Warming Feed the Warming?* ed. George M. Woodwell and Fred T. Mackenzie (Oxford: Oxford University Press, 1995). See also Ben Bond-Lamberty and Allison Thomson, "Temperature-Associated Increases in the Global Soil Respiration Record," *Nature* 464 (March 25, 2010): 579–582; C. Tarnocai, J. Kimble, and G. Broll, "Determining Carbon Stocks in Cryosols Using the Northern and Mid Latitudes Soil Database," in *Permafrost: Proceedings of the 8th International Conference on Permafrost, Zurich, Switzerland, 21–25 July 2003* (London: Momenta, 2003), 1129–1134; J. G. Bockheim, "The Importance of Cryoturbation in Redistributing Organic Carbon in Permafrost-Affected Soils," *Soil Science Society of American* 71, no. 4 (July 2007): 1889–1892;. C. Tarnocai, J. G. Canadell, E. A. G. Schuur, P. Kuhry, G. Mazhitova,

and S. Zimov, "Soil Organic Carbon Pools in the Northern Circumpolar Permafrost Region," *Global Biogeochemical Cycles* 23 (2009).

16. National Research Council, *Abrupt Impacts of Climate Change: Anticipating Surprises* (Washington, DC: National Academies Press, 2013); AAAS, *What We Know: The Reality, Risks, and Response to Climate Change* (Washington, DC: AAAS).

17. Gregory White, *Climate Change and Migration: Security and Borders in a Warming World* (Oxford: Oxford University Press, 2011).

18. IPCC Fourth Assessment Report "Historic Variations in Sea Levels, Part 1: From the Holocene to Romans," http://curryja.files.wordpress.com/2011/07/document. pdf (accessed June 8, 2015). See also Richard S. Williams Jr. and Jane G. Ferrigno, eds., "Satellite Image Atlas of Glaciers of the World" (professional paper 1386, US Geolological Survey, 2012).

19. John F. Hoffecker, Scott A. Elias, and Dennis H. O'Rourke, "Out of Beringia?" *Science* 343, no. 6174 (February 28, 2014): 979–980.

20. National Climate Assessment, *Sea Level Rise* (Washington, DC: US Global Change Research Program), http://nca2014.globalchange.gov/report/our-changing-climate/ sea-level-rise (accessed April 2015).

21. Jet Propulsion Laboratory, "Greenland Ice Loss Doubles in Past Decade, Raising Sea Level Faster," Pasadena, CA, February 16, 2006. See also NEEM Community Members, "Eemian Interglacial Reconstructed from a Greenland Folded Ice Core," *Nature* 493 (January 24, 2013): 489–494.

22. Orrin H. Pilkey and Keith C. Pilkey, *Global Climate Change: A Primer* (Durham, NC: Duke University Press, 2011); Orrin H. Pilkey, William J. Neal, Joseph T. Kelley, and J. Andrew G. Cooper, *The World's Beaches: A Global Guide to the Science of Shoreline*; (Berkeley: University of California Press, 2011); Orrin H. Pilkey and Rob Young, *The Rising Sea* (Washington, DC: Island Press, 2011); Orrin H. Pilkey and Linda Pilkey-Jarvis, *Useless Arithmetic: Why Environmental Scientists Can't Predict the Future* (New York: Columbia University Press, 2007); William J. Neal, Orrin H. Pilkey, and Joseph T. Kelley, *Atlantic Coast Beaches: A Guide to Ripples, Dunes, and Other Natural Features of the Seashore* (Missoula, MT: Mountain Press Publishing, 2007); Orrin H. Pilkey, *A Celebration of the World's Barrier Islands* (New York: Columbia University Press, 2003); Orrin H. Pilkey, Tracy Monegan Rice, and William J. Neal, *How to Read a North Carolina Beach: Bubble Holes, Barking Sands, and Rippled Runnels* (Chapel Hill: University of North Carolina Press, 2004); Orrin H. Pilkey and Katharine Dixon Wheeler, *The Corps and the Shore* (Washington, DC: Island Press, 1996).

23. Robert E. Keane, D. F. Tomback, C. A Aubry, A. D. Bower, E. M. Campbell, C. L Cripps, M. B. Jenkins et al. A Range-Wide Restoration Strategy for Whitebark Pine *(Pinus albicaulis),"* US Department of Agriculture, general technical report, June 2012.

24. Janet Larsen, "Setting the Record Straight: More than 52,000 Europeans Died from Heat in Summer 2003," Plan B Updates, July 28, 2006 (Washington, DC: Earth Policy Institute, 2006).

25. Jeff Masters, "Quiet Tropics—for Now," *Weather Underground* (blog), July 26, 2006, http://www.wunderground.com/blog/JeffMasters/archive.html?year=2006& month=07 (accessed August 20, 2015).

26. Robert Repetto, *The Climate Crisis and the Adaptation Myth* (New Haven, CT: Yale School of Forestry and Environmental Studies, 2008).

27. Shaobing Peng, Jianliang Huang, John E. Sheehy, Rebecca C. Laza, Romeo M. Visperas, Xuhua Zhong, Grace S. Centeno et al., "Rice yields decline with higher night temperature from global warming." *PNAS* 101 (2004): 27: 9971–9975.

CHAPTER 8: THE LIMITS OF BIODIVERSITY

1. The most serious smog event in London, possibly the worst until current problems in China, occurred in December 1952, when several days of temperature inversion held coal smoke close to the ground and caused the death of thousands. Similar problems emerged in Pittsburgh subsequently and on occasion in many other industrial regions. See Devra Davis, *When Smoke Ran Like Water* (New York: Basic Books, 2003).

2. See various authors in E. O. Wilson, ed., *Biodiversity* (Washington, DC: National Academy Press, 1988). See also Norman Myers, "The Biodiversity Challenge: Expanded Hot-Spots Analysis," *Environmentalist* 10, no. 4 (1990): 243–256.

3. Richard O. Bierregaard Jr., Claude Gascon, Thomas E. Lovejoy, and Rita Mesquita, eds., *Lessons from Amazonia: The Ecology and Conservation of a Fragmented Forest* (New Haven, CT: Yale University Press, 2001). See also Jeff Tollefson, "Splinters of the Amazon," *Nature* 496 (2013): 286–289.

4. This topic has been examined in detail in the past from various angles including especially the obvious relationship between the number of species and size of the area considered—the species/area curve. Such analyses turn quickly to explorations of environmental gradients and the changes in communities associated with them. See, for instance, Robert H. Whittaker, *Communities and Ecosystems*, 2nd ed. (New York: Macmillan, 1975). More recently, various experimental approaches have looked at the relationships with diversity in a practical context. See David Tilman and John Downing, "Biodiversity and Stability in Grasslands," *Nature* 6461 (1994): 363–365; David Tilman, "Competition and Biodiversity in Spatially Structured Habitats," *Ecology* 75 (1994): 2–16.

5. National Research Council, *Biological Survey for the Nation* (Washington, DC: National Academy Press, 1993).

6. Endangered Species Act of 1973 (16 U.S.C. 1531–1544, 87 Stat. 884), as amended—Public Law 93–205, approved December 28, 1973, repealed the Endangered Species Conservation Act of December 5, 1969 (P.L. 91–135, 83 Stat. 275). The 1969 act had amended the Endangered Species Preservation Act of October 15, 1966 (P.L. 89–669, 80 Stat. 926). The 1973 act implemented the Convention on International Trade in Endangered Species of Wild Fauna and Flora (T.I.A.S. 8249), signed by the United States on March 3, 1973, and the Convention on Nature Protection and Wildlife Preservation in the Western Hemisphere (50 Stat. 1354), signed by the United States on October 12, 1940. See also Bruce Babbitt, *Cities in the Wilderness: A New Vision of Land Use in America* (New York: Island Press, 2005) 60, 74–75. Babbitt, as secretary of the interior, recognized the unique power of the ESA in giving government powers over the management of public and private land and water in the public interest.

7. Elizabeth Kolbert, *The Sixth Extinction* (New York: Henry Holt, 2014).

8. Mark Kurlansky, *Cod: A Biography of a Fish That Changed the World* (Boston: Addison-Wesley, 1997).

9. I know of no direct study of these populations proving them distinct. I make the suggestion on the basis of numerous analogous studies of other plant and animal populations. We draw on such experience regularly in managing biotic resources. Foresters, for example, pay close attention to "provenance" in selecting seedlings for plantations. They want seedlings from populations that thrive in similar climates and soils. Virtually all populations are ecotypes, some more narrowly restricted than others. The inshore environment is substantially different from the deeper offshore waters, and the life cycles of populations of cod on the two sites were undoubtedly different. The fact that the inshore population remains low to nonexistent suggests little or no exchange between the two.

10. Gretchen G. Daily and Katherine Ellison, *The New Economy of Nature: The Quest to Make Conservation Profitable* (Washington, DC: Island Press, 2007). See also Paul Hawken, Amory Lovins, and L. Hunter Lovins, *Natural Capitalism: The Next Industrial Revolution* (London: Earthscan, 1999). Discussion adapted from George M. Woodwell, "The Biodiversity Blunder," *BioScience* 60, no. 11 (2010): 870–871.

CHAPTER 9: GOVERNMENTS

1. Philippe Girard, *Haiti* (New York: Palgrave Macmillan, 2005). See also George M. Woodwell, *The Nature of a House* (Washington, DC: Island Press, 2009).

2. A particularly destructive Supreme Court decision in a case known as *Citizens United v. Federal Elections Commission* (January 21, 2010) allowed corporations to spend large amounts in election campaigns, thereby giving them substantial political power.

3. Population Reference Bureau, "World Population Data Sheet 2013," 2013, http://www.prb.org/Publications/Datasheets/2013/2013-world-population-data-sheet.aspx (accessed June 9, 2015).

4. For small numbers, the number of interactions rises as $n(n - 1)$. For larger numbers, above about 5, the significance of the minus one is of no consequence. Hence it is known as the N-squared law.

5. Derrick Jensen, *Endgame, Vol. 1: The Problem of Civilization* (New York: Seven Stories Press, 2006).

6. Sheldon S. Wolin, "Inverted Totalitarianism," *Nation*, May 19, 2003; Sheldon S. Wolin, *Politics and Vision: Continuity and Innovation in Western Political Thought*, 2nd ed. (Princeton, NJ: Princeton University Press, 2004).

7. C.A.S. Hall and his colleagues at Syracuse have examined the net cost of energy development and expressed it in units of energy. A crude estimate by Hall of the cost of tar sands oil is about one-quarter barrel per barrel of refined oil (telephone conversation, May 2014).

8. Craig Simons, *The Devouring Dragon* (New York: St. Martin's Press, 2013). Based on years of experience in China, Simons presents a disturbing view of the expansion of a new corporate and industrial, all-consuming giant. See also John Cassidy, "Is China the Next Lehman Brothers?" *New Yorker*, April 3, 2014, http://www.newyorker.com/rational-irrationality/is-china-the-next-lehman-brothers (accessed June 9, 2015).

9. Chris Hedges and Joe Sacco, *Days of Destruction, Days of Revolt* (New York: Nation Books, 2012).

10. In January 2014, the temperatures in central Australia exceeded 120 degrees Fahrenheit, killing animals and threatening humans who could not escape exposure. See Will Oremus, "Australia Is So Hot They Had to Add a New Color to the Weather Map," http://www.slate.com/blogs/future_tense/2013/01/08/australia_heat_wave_new_color_added_to_weather_maps_fire_danger_catastrophic.html (accessed August 20, 2015).

11. Paul Hawken, *The Ecology of Commerce: A Declaration of Sustainability* (New York: HarperCollins, 1993); William McDonough and Michael Braungart, *Cradle to Cradle* (New York: North Point Press, 2002).

12. Eric T. Freyfogle, *The Land We Share: Private Property and the Common Good* (Washington, DC: Island Press, 2003).

13. Nicola Clark, "State Tax Deal for Boeing Draws an EU Challenge," *New York Times*, December 19, 2014. http://www.nytimes.com/2014/12/20/business/international/european-union-boeing-subsidies-wto.html?ref=topic (accessed August 20, 2015).

14. Editorial Board, "Fitful Progress in the Antismoking Wars," *New York Times*, January 9, 2014, http://www.nytimes.com/2014/01/10/opinion/fitful-progress-in-the-antismoking-wars.html (accessed June 9, 2015).

15. David Goldstein, "The Low-Hanging Fruit … That Keeps Growing Back," Switchboard, Natural Resources Defense Council, August 26, 2011, http://switchboard.nrdc.org/blogs/dgoldstein/post_1.html (accessed June 9, 2015). Goldstein shows how governmental regulations have resulted in continuous improvements in refrigeration, thus saving the public billions in costs as well as many tons of fossil fuels and associated pollution over decades.

16. Peter Lehner with Bob Deans, *In Deep Water: The Anatomy of a Disaster, the Fate of the Gulf, and Ending Our Oil Addiction* (New York: Natural Resources Defense Council, 2011).

17. Edward Broughton, "The Bhopal Disaster and Its Aftermath: A Review," *Environmental Health* 4, no. 6 (2005), http://www.ncbi.nlm.nih.gov/pmc/articles/PMC1142333/ (accessed June 10, 2015).

18. Nick Surgey, "Revealed: ALEC's 2014 Attacks on the Environment," *Huffington Post*, May 5, 2014, http://www.huffingtonpost.com/nick-surgey/revealed-alecs-2014-attac_b_5256001.html (accessed June 10, 2015).

19. Mae Wu, Mayra Quirindongo, Jennifer Sass, and Andrew Wetzler, *Poisoning the Well: How the EPA Is Ignoring Atropine Contamination in Surface and Drinking Water in the Central United States* (New York: NRDC Special Publication, 2009).

20. Abraham Lincoln, Gettysburg Address, 1863.

CHAPTER 10: A NEW DEPARTURE

1. James Gustave Speth and Peter M. Haas, *Global Environmental Governance:* (Washington, DC: Island Press, 2006).

2. "New Delhi Declaration of Principles of International Law Relating to Sustainable Development," *International Environmental Agreements: Politics, Law, and Economics* 2 (April 2, 2002): 211–216.

3. Nancy N. Rabalais, R. Eugene Turner, and William J. Wiseman Jr., "Gulf of Mexico Hypoxia, aka 'The Dead Zone,'" *Annual Review of Ecology and Systematics* 33 (2002): 235–263. Nancy N. Rabalais, R. Eugene Turner, and Donald Scavia, "Beyond Science into Policy: Gulf of Mexico Hypoxia and the Mississippi River," *BioScience* 52, no. 2 (2002): 129–142.

4. International Survey of Herbicide Resistant Weeds, http://www.weedscience.org/References/ReferenceView.aspx (accessed April 10, 2015).

5. Frederick N. Fishel, "Pesticide Use Trends in the U.S.: Global Comparison," University of Florida at Gainesville, IFAS Extension, 2007, revised 2013.

6. Cheryl Lyn Dybas, "Polar Bears Are in Trouble—and Ice Melt's Not the Half of It," *BioScience* 62, no. 12 (2012): 1014–1018.

7. Frederica P. Perera and Julie Herbstman, "Prenatal Environmental Exposures, Epigenetics, and Disease," *Reproductive Toxicology* 31, no. 3 (April 2011): 363–373.

8. Herman Daly, *Beyond Growth: The Economics of Sustainable Development* (Boston: Beacon Press, 1996).

9. Since 1934, the US Department of Agriculture has maintained a series of experimental watersheds in Otto, North Carolina, to provide basic information about the management of drainage basins under various regimes. Since the early 1960s on Forest Service land near Plymouth, New Hampshire, scholars from around the world have conducted a series of studies in the Hubbard Brook drainage to build an even larger body of data on the role of forests in biogeochemical cycles. The initiative was taken and supported over the ensuing decades by F. Herbert Bormann and Gene E. Likens, both then at Dartmouth College.

10. Food and Agriculture Organization of the United Nations, "Global Forest Resources Assessment 2010." See also World Commission on Forests and Sustainable Development, *Our Forests, Our Future: Report of the World Commission on Forests and Sustainable Development* (Cambridge: Cambridge University Press, 1999).

11. Taryn Fransen, Mengpin Ge, and Thomas Damassa, "The China-U.S. Agreement: By the Numbers," World Resources Institute, November 20, 2014, http://www.wri.org/blog/2014/11/numbers-china-us-climate-agreement (accessed June 10, 2015).

12. Yude Pan, Richard A. Birdsey, Jingyun Fang, Richard Houghton, Pekka E. Kauppi, Werner A. Kurz, Oliver L. Phillips et al., "A Large and Persistent Carbon Sink in the World's Forests," *Science* 333 (2011): 988–993.

13. George M. Woodwell, R. A. Houghton, Eric A. Davidson, and Daniel C. Nepstad, "The First Principles for Climatic Stabilization," *Carbon Management* 2 (2011): 605–606. See also Pierre Friedlingstein et al., "Update on CO2 Emissions," *Nature Geoscience* 3 (2010): 811–812. See also World Commission on Forests and Sustainable Development, *Our Forests, Our Future.*

14. Committee on Understanding and Monitoring Abrupt Climate Change and Its Impacts, Board on Atmospheric Sciences and Climate, Division on Earth and Life Studies, and National Research Council, *Abrupt Impacts of Climate Change: Anticipating Surprises* (Washington, DC: National Academies Press, 2013). See also G. Hugelius et al., "The Northern Circumpolar Soil Carbon Database: Spatially Distributed Datasets of Soil Coverage and Soil Carbon Storage in the Northern Permafrost Regions," *Earth System Science Data* 5 (January 5, 2013): 3–13, http://www.earth-syst-sci-data.net/5/3/2013/essd-5-3-2013.html (accessed June 10, 2015).

15. George M. Woodwell and Fred T. Mackenzie, eds., *Biotic Feedbacks in the Global Climatic System* (Oxford: Oxford University Press, 1995); Christopher Stil, "As Different as Night and Day," *Nature* 501 (September 5, 2013): 39–40; Shushi Peng, Shilong Piao, Philippe Ciais, Ranga B. Myneni, Anping Chen, Frédéric Chevallier, Albertus J. Dolman et al., "Asymmetric Effects of Daytime and Night-Time Warming on Northern Hemisphere Vegetation," *Nature* 501 (September 5, 2013): 88–92.

16. Thomas F. Stocker, "The Closing Door of Climate Targets," *Science* 339 (2013): 280–282.

17. A. Baccini et al., "Estimated Carbon Dioxide Emissions from Tropical Deforestation Improved by Carbon-Density Maps," *Nature Climate Change* 2 (2012): 182–185; S. J. Goetz et al., "Observations and Assessment of Forest Carbon Dynamics Following Disturbance in North America," *Journal of Geophysical Research* 117 (2012), doi: 10.1038/nclimate1354.

18. Richard A. Houghton, "The Emissions of Carbon from Deforestation and Degradation in the Tropics: Past Trends and Future Potential," *Carbon Management* 4, no. 5 (2013): 539–546.

CHAPTER 11: *SIC UTERE*

1. Paul R. Ehrlich, *The Population Bomb* (New York: Sierra Club-Ballantine Books, 1968); Paul R. and Anne H. Ehrlich, "What about the Population Bomb?" in *Ecology and the Common Good: Great Issues of Environment*, ed. Richard A. Houghton and Allison B. White (Falmouth, MA: Woods Hole Research Center, 2014).

2. Lester R. Brown, *World on the Edge: How to Prevent Environmental and Economic Collapse* (New York: W. W. Norton, 2011).

3. John Vidal, "Corporate Stranglehold of Farmland a Risk to World Food Security, Study Says," *Guardian*, May 28, 2014, http://www.theguardian.com/environment/2014/may/28/farmland-food-security-small-farmers (accessed June 11, 2015).

4. *Bloomberg News*, "Death in Parched Farm Field Reveals Growing India Water Tragedy," May 21, 2013, http://www.bloomberg.com/news/articles/2013-05-21/death-in-parched-farm-field-reveals-growing-india-water-tragedy (accessed October 21, 2015).

5. Lester R. Brown, *Outgrowing the Earth: The Food Security Challenge in an Age of Falling Water Tables and Rising Temperatures* (New York: W. W. Norton, 2004); Lester R. Brown, *Plan B 3.0: Mobilizing to Save Civilization* (New York: W. W. Norton, 2008).

6. Ian Scott-Kilvert, trans., *Makers of Rome: Nine Lives by Plutarch* (New York: Penguin Books, 1965).

7. Robert Kuttner, "Karl Polanyi Explains It All," *American Prospect*, April 15, 2014, http://prospect.org/article/karl-polanyi-explains-it-all (accessed June 11, 2015).

8. Thomas Piketty, *Capital in the Twenty-First Century* (Cambridge, MA: Harvard University Press, 2014); Elizabeth Warren, *A Fighting Chance* (Boston: Macmillan, 2014).9. Herman E. Daly, "From Empty-World Economics to Full-World Economics: A Historical Turning Point in Economic Development," in *World Forests for the Future: Their Use and Conservation,* ed. Kilaparti Ramakrishna and George M. Woodwell (New Haven, CT: Yale University Press, 1993).

10. James Gustave Speth, *America the Possible: Manifesto for a New Economy* (New Haven, CT: Yale University Press, 2012); William Mc Donough and Michael Braungart, *Cradle to Cradle: Remaking the Way We Make Things* (New York: North Point Press, 2002); James Howard Kunstler, *The Long Emergency: Surviving the End of Oil, Climate Change, and Other Converging Catastrophes of the Twenty-First Century* (New York: Grove Press, 2005); Richard Heinberg, *Power Down* (Gabriola Island, BC: New Society Publishers, 2004).

11. Ron Nixon, "Senate Passes Long-Stalled Farm Bill, with Clear Winners and Losers," *New York Times,* February 4, 2014, http://www.nytimes.com/2014/02/05/us/politics/senate-passes-long-stalled-farm-bill.html (accessed June 11, 2015).

12. Jared Diamond, *Collapse: How Societies Choose to Fail or Succeed* (New York: Viking, 2005).

13. Robert Repetto, *World Enough and Time: Successful Strategies for Resource Management* (New Haven, CT: Yale University Press, 1986).

14. Wes Jackson, *Consulting the Genius of the Place: An Ecological Approach to a New Agriculture* (Berkeley, CA: Counterpoint Publishers, 2010); Tom MacMillan and Tim G. Benton, "Engage Farmers in Research," *Nature* 509 (May 2014): 25–27. The group in Salina has developed a perennial strain of wheat they call "Kernza," which is being tested elsewhere commercially; see Land Report, no. 109, summer 2014, Land Institute, http://www.landinstitute.org/wp-content/uploads/2014/12/LR-109.pdf (June 11, 2015). See also Andrew W. Pollack, "Genetic Weapon against Insects Raises Hope and Fear in Farming," *New York Times,* January 28, 2014, A3; Andrew Warren, Thomas E. Gill, and John E. Stout, "The Bibliography of Aeolian Research," Cropping Systems Research Laboratory, US Department of Agriculture, Lubbock, TX, 2008.

15. GRAIN, "Hungry for Land: Small Farmers Feed the World—with Less than a Quarter of All Farm Land," May 28, 2014, http://www.grain.org/article/entries/4929-hungry-for-land-small-farmers-feed-the-world-with-less-than-a-quarter-of-all-farmland (accessed June 11, 2015). See also Klaus Deininger and Derek Byerlee, *Rising Global Interest in Farmland: Can It Yield Sustainable and Equitable Benefits?* (Washington, DC: World Bank, 2011).

16. Ambrose Bierce, *The Devil's Dictionary* (1911). Bierce published several lists of humorous definitions over more than twenty years after 1880.

17. Jackson, *Consulting the Genius of the Place*.

18. Paul Hawken, *The Ecology of Commerce: A Declaration of Sustainability* (New York: HarperCollins, 1994).

19. McDonough and Braungart, *Cradle to Cradle*.

20. Linda Greer, "Preventing Industrial Pollution at Its Source: A Final Report of the Michigan Source Reduction Initiative," Natural Resources Defense Council, 1999, http://www.nrdc.org/water/pollution/msri/msriinx.asp (accessed June 11, 2015).

21. Elana Mass, "Watershed Protection and New York City's Water Supply," Prince William Conservation Alliance, Woodbridge, VA, http://www.pwconserve.org/issues/watersheds/newyorkcity (accessed June 11, 2015); City of New York, "History of New York City's Water Supply System," NYC Environmental Protection, http://www.nyc.gov/html/dep/html/drinking_water/history.shtml (accessed June 11, 2015); Elizabeth Royte, "On the Waterfront," *New York Times*, February 18, 2007.

22. Kuttner, "Karl Polanyi Explains It All," 70–75.

23. Henry M. Paulson, "The Coming Climate Crash: Lessons for Climate Change in the 2008 Recession," *New York Times*, Sunday Review, June 21, 2014.

24. The potential for producing electric energy locally has already reached into far corners. Travel by boat on the rivers of the Amazon in Brazil and Peru reveals there is scarcely a village that does not have a solar panel on a thatched roof near the village center powering a radio, telephone, or computer. Such local potential in producing electric energy opens a range of further developments, all consistent with gradually building locally focused life-support systems that introduce new potential in self-sufficiency within the matrix of the natural systems we all inhabit.

Accumulation, of toxins, 23, 26–27, 31–33, 36, 39–41, 46
Adams, John, vii
Adaptation myth, xii, 101, 104–118. *See also* Biosphere, resilience of, as misconception
Adirondack Park, New York, 189
ADM, 178
Africa, 53–54, 77, 80
African violets, 16
Agriculture. *See also* Industrial agriculture; University agriculture departments/schools
  climate change's effect on, 84, 117
  sustainable model of, 180–182
  tropical forest destruction by, 49–52
Air pollution, 158
Alberta tar sands, 70, 78, 141, 143, 144
Alfven, Hannes, 25–26
Alvarez, Robert, 23–24
Amazon basin, 49–51, 62–63, 71–72, 84, 125, 160
America Farm Bureau Federation, 178
American Legislative Exchange Council, 100, 152
American Petroleum Institute, 152
Antarctic region, 61, 113
Arctic region, 18, 40, 58–61, 98, 104, 106, 108, 110, 158, 166
Arrhenius, Svante, 90
Atomic Energy Commission, 14, 16
Ayres, Richard, viii

Baird, Spencer, 130
Bee colony collapses, 38
Bhopal chemical accident, 150
Bierce, Ambrose, 182
Bikini Atoll, 11
Biodiversity, 119–133
  concept of, 124
  emphasis on, 125–129
  hot spots of, 124–125, 127–129
  research on, 125
  value of, 131–132
Biosphere. *See also* Environmental system
  actions required to nourish, 132–133, 155–171
  chronic disturbance of, 19–20, 26–27, 41, 45, 50, 112, 118, 127–128, 132–133, 157, 174
  as the commons, 27
  degradation of, 42, 76–77, 83–84
  development of, 3–4
  disasters affecting, 7
  effects of ionizing radiation on, 11–19
  importance of place/location in, 120–121, 125–127, 129–131, 168
  public welfare and, 202n15
  resilience of, as misconception, 4–5, 34, 121 (*see also* Adaptation myth)
  sacrificial zones in, 6, 42, 46–47, 144
  as sum of its parts, 174
  as system, 159–160
  tropical forests and, 49–52

Biotic communities, 40–41, 159–161
Bittersweet (plant), 160
Boeing, 147–148
Boreal forests, 19, 29–30, 50, 58, 59,
    104, 106, 109–110, 121, 124, 166
Bormann, F. H., 16
BP (British Petroleum), 42–43, 74, 137,
    149–150
Bradley, David, 12
Braungart, Michael, 177, 183
Brookhaven National Laboratory, 14,
    16–18, 35, 55–58
Brown, Lester, vii, 9, 79, 81–82, 174–175
Bush, George H. W., 152
Bush, George W., 98–99

California, 84, 92
Canada, 78, 143
Cap-and-trade programs, 92, 132, 191
Carbon, 51–63, 110, 161, 163–166, 168,
    170–171, 180–181, 192, 201n11
Carbon dioxide, 17, 52–57, 61–62, 90,
    103–104, 109, 163–164, 171, 201n7
Cargill, 178
Carson, Rachel, 11, 33
    Silent Spring, x, 33–35, 157
Carter, Jimmy, 90–92, 152
Cesium, 17–18, 23
Chelyabinsk, Russia, 21
Chernobyl nuclear accident, 21
Chevron, 71–72, 74
China
    climate change's effect on, 53, 112
    environmental policies of, 143, 162
    water shortages in, 175
Chronic disturbance, of biosphere,
    19–20, 26–27, 41, 45, 50, 112, 118,
    127–128, 132–133, 157, 174
Citizens United v. Federal Elections Com-
    mission (2010), 77, 176, 211n2
Civil rights
    as common property, 76, 86
    evolution of, 3

industrial economy balanced with,
    140
    protection of, 86–87
    struggles over, 76
Clean Air Act, 151
Climate change
    actions required to address, 61–62, 95–
        96, 99–101, 103–104, 106, 117–118,
        161–171
    conclusions from research on, 59–60
    denial or ignoring of, 8, 57, 73–74,
        100, 106–107, 111, 118, 187
    economic effects of, 116–117
    effects of, 7–9, 20, 53–54, 60–61, 95–
        96, 98, 99, 103–104, 106, 109–118
    feedback problem concerning,
        108–111
    forests and, 56–57
    fossil fuels and, 59, 76–77, 90, 92,
        94–95, 100, 104, 107
    international treaty on, 96–97
    mitigation of, 105
    politics and, 89–101, 106–107
    research on, 54–58, 90, 103
    reversal of, 106, 109, 111, 161–162,
        167, 169–170
    tropical forests and, 49–51
Closed-cycle industry, 24
    as closed system, xiii, 13, 27, 46, 174,
        186
Cloud, Preston, 29
Coal, 26, 79, 106, 162
Cochran, Thomas, 22, 27
Cod, 129–131
Columbus, Christopher, 80
Common Cause, 92
Commons, global, xiii, 67–83. See also
    Tragedy of the commons
Conference of the Parties to the Frame-
    work Convention on Climate
    Change, 97, 98, 103, 162
Conservation, 122–129, 153
Conservation International, 188

Contamination. *See also* Toxins
  of air, 158
  corporate prevention of, 183–186
  DDT, 30–32
  examples of, 39–40
  of food webs, 31
  of global cycles, 158
  herbicidal, 26, 38
  nuclear, 12–14, 21–23, 32, 34, 72–73
Convention on International Trade in
    Endangered Species, 123, 211n6
Conway, Eric, 68
Corporations. *See also* Industry
  and conservation movement, 128
  demands placed on resources by, ix–xi
  ecological damage caused by, 5, 6, 8,
    173
  government relations with, 5, 8, 32–
    33, 36–37, 141–143, 145, 149–150,
    152
  as individuals, 138–140, 148
  power and influence of, ix, 6, 8, 77–
    79, 98, 147, 176, 211n2
  public subsidies for, 44–45, 74, 143–
    144, 147–148
  responsibilities of, 145–146, 148,
    183–187, 192
  rights of, 138–140, 148
  science biased toward, 68–69
  self-serving behavior of, 75, 149, 152
Correa, Rafael, 78
Council on Environmental Quality,
    90–91, 151
Cutiar, Michael Zammit, 97

Daly, Herman, 83–84, 149, 177
Darwin, Charles, xii, 3–5, 10, 33, 156
DDT, 29–41
  alternatives to, 38
  distortions concerning, 68
  effectiveness of, 197n5
  effects of, 30–32
  history of, 33

lessons learned from, 46
  mechanism of transfer, 31
  persistence of, 34
  reasons for using, 30–33, 36
  regulation of, 35–36
  spraying methods, 30, 31
Decay, 51, 58
Declaration of Independence, 138,
    153–154
Democratic capitalism, 69, 77, 120, 138,
    173
Diamond, Jared, 179
Dose, toxicity in relation to, 34, 36, 46
Dow Chemical, 183–186
Drainage basins, 82, 85, 86, 113, 125,
    126, 149, 160, 166, 168, 187, 190
Drought, 9, 112

Earth. *See* Biosphere
Earth Summit (Rio de Janeiro, 1992),
    96–97
Ecology
  media and, 70
  oppositional viewpoints, 8, 34–37
  roots of, 3–5, 156
Economic growth, ix–x, 75, 121, 159
Economic system, 6, 9–10, 83–85, 173
Ecotypes, 130, 131
Ecuador, 71–72, 78
Ehrlich, Paul, viii
  The Population Bomb, x
Emerald ash borer, 160
Endangered Species Act (ESA), 123, 125–
    127, 211n6
Endocrine system, 41
Energy production
  local, 24, 25, 217n24
  renewable, 24, 25, 92
Eniwetok, 11
Environment. *See* Biosphere
Environmental Defense Fund, 92, 153,
    188
Environmental Impact Statement, 93

Environmental Protection Agency (EPA)
and DDT, 35
establishment of, 37, 152
petition and review process of, 38
phasing out of coal by, 106, 162
Environmental system, 6, 9–10, 83–85, 173. *See also* Biosphere
EPA. *See* Environmental Protection Agency
ESA. *See* Endangered Species Act
European Union (EU), 42
Evaporation, 49–50
Exxon, 98
ExxonMobil, 8, 107

Farm Bill (US), 177
Farmer, Paul, 82
FCCC. *See* Framework Convention on Climate Change
Federal Crop Insurance Program, 117
Feedback problem, 108–111
Field-to-forest succession, 19
Fishing, commercial, 129–131
Food web contamination, 30–31
Forbes, Steve, 73
Forests. *See also* Boreal forests; Tropical forests
analytical value of, 160
metabolism of, 17, 56–57, 163, 166
preservation and restoration of, 165–169
significance of, in ecological management, 163–169
Fossil fuel industry
abandonment of, 170–171, 191–192
and climate change, 53, 59, 76–77, 90, 92, 94–95, 100, 104, 107, 117–118
debt owed to nature by, 180
economics of, 78
harms resulting from, xii, 7, 42, 47, 49, 78–79
power and influence of, 8, 100, 141, 143

public subsidies for, 44, 143–144
Framework Convention on Climate Change (FCCC), 97–98, 150, 164, 167, 180, 187
Conference of the Parties, 97, 98, 103, 162
France, 115
Free market, 69, 70, 73, 82, 107, 151, 191
Fukushima, Japan, 21–22, 46, 70–72, 137

Gaia hypothesis, x
General Motors, 75
Genetic engineering, 146
Geoengineering, 62
Georges Bank, 113, 131
Global commons, xiii, 67–83
Golden rule, 85, 146, 156, 173
Goldsboro, North Carolina, 206n1
Gordon, Alan, 19
Gordon, James, 107
Gore, Al, 69
Gorham, Eville, 19
Government, 137–154. *See also* Government regulation
corporate relations with, 5, 8, 32–33, 36–37, 141–143, 145, 149–150, 152, 211n2
ecological disasters addressed by, 141–142
popular attitudes toward, 73, 148
protective function of, 7, 8, 39–40, 139–141, 145, 153
role of, 33, 92, 98, 144–145, 148–149
Government regulation, ix
of DDT, 35–36
in EU, 42
laxity of, 36, 42
of nuclear industry, 13, 15, 24, 26–27
opposition to, 73–75
public interest served by, 149
GRAIN, 181

Grand Banks of Newfoundland, 130–131
Greenhouses gases, 45
Greenland, 179
Greenpeace, 188
Greer, Linda, 183–186
Gulf of Mexico, 42–44, 46–47, 74, 137, 149, 157

Haas, Peter, 155–156
Haiti, 80–82, 85–86, 137
Hammond, E. Cuyler, 67
Hansen, James, 24, 94–95
Hardin, Garrett, x, 6–7, 10, 69
Harper, Stephen, 78, 143
Hatteras Island, North Carolina, 115
Hawken, Paul, 183
Hawkins, David, 45
Heat, associated with climate change, 115–116
Hedges, Chris, 6, 42, 144
Helms, Jesse, 97
Herbicides and herbicide-resistant crops, 38, 147, 152–153, 157
Holdren, John, viii, 105
House Subcommittee on Interior and Insular Affairs, 94
Human birthright, x, 35, 40, 86, 108, 138, 145, 147, 156
Human rights, 81, 139, 140, 153–154
Humans. *See also* Population growth
ecological impact of, 4, 12, 43, 120–121
vulnerability of, to toxins, 13, 23, 26, 32, 41, 46, 158
Hurricane Irene, 115
Hurricane Katrina, 60, 108, 110
Hurricane Sandy, 7, 60, 108
Hydrogen, 191

Impoverishment, biotic
from climate change, 62, 115
examples of, 76

of forests, 160
in Haiti, 80, 82–83
incremental, 31, 32
from industrial development, 79
relation of economy to, 83–84
social upheavals triggered by, 112
in Somalia, 54
systematic, 19–20, 26–27, 41, 44–45, 127, 132–133, 169, 174
India, 175
Individual, concept and role of, 148
Industrial agriculture
alternatives to, 180–182
criticisms of, 174–178, 181–182
ecological effects of, 26, 152–153, 157, 174–175
no-till, 38, 147
power and influence of, 177
public subsidies for, 158, 178
social harms caused by, 175
Industrial development, ecological impacts of, ix–x, 6–7, 31, 72, 79
Industry. *See also* Corporations; Industrial development
as closed system, xiii, 13, 24, 27, 46, 174, 186
harms resulting from, 150
limitations on liability of, 14–15, 20–21
Inhofe, James, 8, 100
Intensification, ix
International Union for Conservation of Nature (IUCN), 122–123, 188
Inverted totalitarianism, 79, 98, 142–143
Iodine 131, 23
Islands, 114
IUCN. *See* International Union for Conservation of Nature

Jackson, Wes, 181, 182
Jensen, Derrick, 141
Johnson, Lyndon, 90

Kazakhstan, 99
Keeling, Charles David, 54–56, 90, 91
Koch, William, 107
Kolbert, Elizabeth, 127
Kudzu, 159
Kunstler, James Howard, 177
Kuttner, Robert, 191
Kwajalein, 11
Kyoto Protocol, 97–98, 162
Kyshtym-57 disaster, 21

Land Institute, 181
Land use policies, 189–190
Leopold, Aldo, 173
    A Sand County Almanac, x
Limited Test Ban Treaty, 12
Lincoln, Abraham, 3, 10, 89
Lovejoy, Tom, 124–125
Lovelock, James, x
Lovins, Amory, vii

MacDonald, Gordon, 91
MacLeish, Archibald, 49
Malaria, 34, 36
Malthus, Thomas Robert, 9
Mann, Michael, 100
Marsh, George Perkins, xii, 4, 10, 33,
    83, 156
Marshall Islands, 11
Marshes, 41
Masters, Jeff, 116
Mathews, Jessica Tuchman, 94
Mayak nuclear accident, 21
McDonough, William, 177, 183
McKibben, Bill, 61
McMichael, Anthony J., 8
Medvedev, Zhores, 21
Mercury, 40–41
Methane, 53, 58, 61
Mexico, 53, 112
Mexico, Maine, 138–139
Mitigation, of climate disruption, 105
Monsanto, 178

Muller, Paul H., 33
Mutations, 15–16, 19, 20, 195n5

National Academy of Sciences, 37
National Biological Survey, 126, 129
National Commission on the BP Deep-
    water Horizon Oil Spill and Offshore
    Drilling, 149–150
National Environmental Policy Act, 93,
    152
National Research Council, 126
Nations, impoverishment of, 79–80
Natural Resources Defense Council
    (NRDC), 22, 37, 91, 92–93, 153,
    183–186, 188
Nature. See Biosphere
Nature Conservancy, 188
Neonicotinoids, 38
Nepstad, Daniel, 49
Net ecosystem production, 57
Nevada Test Site, 11
New Delhi Declaration of Principles of
    International Law Relating to Sus-
    tainable Development, 156
Newfoundland, 130–131
New Orleans, Louisiana, 60, 108,
    110–111
New York City, New York, 190
New York Times (newspaper), 49–51,
    62–63, 70, 78, 99
Nicotinoids, 38
Nitrogen, 41, 157
Nitrous oxide, 53
Nixon, Richard, 37, 151
Nonprofit sector, 37, 152
Norway, 189
No-till agriculture, 38, 147
NRDC. See Natural Resources Defense
    Council
N-squared law, 140, 174, 191
Nuclear energy, 11–27
    disasters involving, 21–23, 70–72

ecological insights occasioned by, 12, 27
industry liability concerning, 14–15, 20–21
optimism about, 13, 14, 24–25, 119–120
reactor management issues, 23–24
safety concerns about, 13–16, 20–27
Nuclear tests, 11–12
Nuclear weapons, 206n1
effects of, 11–13
proliferation of, 89
safety concerns about, 25–26, 206n1

Oak Ridge National Laboratory, 20, 21
Obama, Barack, 78, 105–106, 162
Oceans, 108–109, 161, 164, 170
Ocracoke Island, North Carolina, 114
Odum, Eugene P., 16
Oeschger, Hans, 55, 201n7
Office of Technology Assessment, 74
Oil disasters, 42–43
Oreskes, Naomi, 68
Organic soils, 180–181
Orr, David, 69
Osborn, Fairfield, *Our Plundered Planet*, x
Ospreys, 30–31

Pakistan, 99
Parks and reserves, 128–129
Passenger pigeons, 122
Paulson, Henry M., 191–192
Pebble Mine, 93
Peregrine falcons, 30
Perera, Frederica, 158
Pesticides, 157–158. *See also* DDT
Photosynthesis, 17, 56–58, 168
Piketty, Thomas, 177
Place, importance of, in biosphere, 120–121, 125–127, 129–131, 168
Plants, radioactive sensitivity of, 14–19
Plowshare, 11
Plutarch, 176

Poisons. *See* Toxins
Polanyi, Karl, 191
Polar bears, 158
Political system, 5, 9–10, 83–85, 173
Politics
and climate change, 89–101, 106–107
corporate influence in, 77–79, 147, 176, 211n2
Pollution. *See* Contamination; Toxins
Population growth, ix–x, 9, 140, 176–177, 191
President's Science Advisory Committee, 90
Price-Anderson Act, 15, 21, 42, 119
Proctor, Robert, 68
Public interest/welfare
biodiversity as contribution to, 131–132
climate change and, 114
the commons as, x–xi, 35
corporate exploitation of, 79, 108, 147, 149–150
environmental aspect of, 202n15
EPA and, 37
government regulation promoting, 149
government responsibility for, 32, 98
non-corporate advocates for, 39
and nuclear concerns, 14
promotion of, 86–87
tobacco industry and, 68
Public subsidies, 44–45, 74, 143–144, 147–148, 158, 178
Putin, Vladimir, 98

Radiation
effects of, 11–19, 22–23, 195n5
safety concerns about, 13–16, 20–27, 45–46
Rain forests. *See* Tropical forests
Ramakrishna, Kilaparti, 96–97
Reagan, Ronald, 73, 75, 91–92, 94, 149, 152

Red List, 122–124
Regional Greenhouse Gas Initiative, 92, 132
Regulation. *See* Government regulation
Remote-sensing techniques, 167
Renewable energy, 24, 92, 107
Repetto, Robert, 116–117, 180
Republican Party, 8, 92, 98
Reserves. *See* Parks and reserves
Respiration, plant, 17, 56–58
Revelle, Roger, 54, 90, 91
Rich-poor gap, 175–177
Rights. *See* Human birthright; Human rights
Rio de Janeiro Earth Summit, 96–97
Rocky Mountain Institute, 92
Roman Empire, 83, 176
Romm, Joe, 8
Rongelap, 11
Rowland, Sherwood, 103
Ruckelshaus, William, 35–36
Russia, 98, 116

Sacco, Joe, 6, 42, 144
Santa Monica, California, 187
Savannah River Laboratory, 16
Science
    corporate distortion of, 68–69
    hostility to, 73–75, 100
    insular focus of, 122, 128
Scientific American (magazine), 203n8
Scripps Institution of Oceanography, 54
Sea levels, 7, 59, 61, 112–114
Sears, Paul, *Deserts on the March*, x
Senate Committee on Energy and Natural Resources, 94
Shakespeare, William, 155
Sic utere principle, 146, 152, 153, 156, 192
Simons, Craig, 143
Small farms, 175, 178, 181
Smog, 122, 210n1
Smoking, 67–68

Somalia, 54, 77, 80
South Africa, 54
Soybeans, 49–51
Sparrow, Arnold, 14–16, 19
Species, preservation of, 122–126, 128–129
Speth, James Gustave, vii, viii, 67, 69, 91, 93–94, 155–156
Spruce budworms, 29–30
Steinbeck, John, *The Grapes of Wrath*, x
Steller's sea cow, 122
Stockholm Environment Institute, 153
Strontium 90, 23
Subsidies. *See* Public subsidies
Subsistence farming, 178–180
Succession, ecological, 19–20
Sulfur oxides, 19
Sweden, 189

Taibbi, Matt, 137
TEPCO, 71
Texaco, 72
Three Mile Island nuclear accident, 21
Tickell, Crispen, 96–97
Tillerson, Rex, 107
Tobacco industry, 67–68
Totalitarianism. *See* Government
Toxins. *See also* Contamination
    accumulation of, 23, 26–27, 31–33, 36, 39–41, 46
    assumption of safety of, 36, 40, 42, 46
    concentrations of, 31, 32
    difficulties surrounding proof concerning, 38–40, 68, 74–75
    distributions of, 31–32, 40
    dosage of, 34, 36, 46
    industrial release of, 36
    persistence of, 34
    zero tolerance for, 35–36, 39, 46
Tragedy of the commons, x–xi, 6–7, 10, 69, 75
Train, Russell, 128
Tropical forests, 49–52, 124, 160
Typhoon Haiyan, 61, 103

Udall, Stewart, 90
Ukraine, 99
Union Carbide, 150
United Nations
    and environmental issues, 96–97
    Environment Programme, 96
    Universal Declaration of Human
        Rights, 139, 145, 154
United States
    climate change's effect on, 53, 98, 99,
        116–117
    and international environmental pol-
        icy, 98–100, 105, 163
    land management in, 189
    socioeconomic conditions in, 6
    water shortages in, 175
Universal Declaration of Human Rights,
    139, 145, 154
University agriculture departments/
    schools, 36–37, 39, 198n8
US Biological Survey, xii
US Congress, xii
    and climate change, 94–95, 98, 100,
        187
    and endangered species, 125–126
    environmental policies of, 151–152
    responsible for ecological harm, 8–9
    shirking of ecological responsibilities
        by, 74
US Department of Agriculture, 189
US Fisheries Commission, 130
US Nuclear Regulatory Commission
    Committee on Nuclear Safeguards,
        22
US Soil Conservation Service, 189

Vaporization, energy resulting from,
    49–50, 103
Vapor phase, 40
Varzea, 62–63
La Via Campesina, 181
Vogt, William, *Road to Survival*, x
Von Hayek, Friedrich, 191

Wallace, Alfred Russel, 4
Warren, Elizabeth, 177
Wastes. *See* Toxins
Water Pollution Control Act, 151
Water supply, 26, 78, 82, 85, 86, 112,
    126, 149, 166, 169, 175, 187, 190.
    See also Drought
Water vapor, 51, 53
Wealth gap, 175–177
Whitman, Walt, 119
Wind farms, 107
Wirth, Tim, 94
Wolin, Sheldon, 79, 142
Woods Hole Research Center, 49, 96–97,
    166
World Bank, 181
World Commission on Forests, 166
World Health Organization, 36
World Resources Institute, 94, 153, 180
Worldwatch Institute, 153
World Wildlife Fund (WWF), 123, 188
WWF. *See* World Wildlife Fund
Wyoming, 26, 70, 79, 141, 144

Yasuni National Park, 78
Yemen, 174

Zero tolerance, for toxins, 35–36, 39, 46